AF522282

Marine Biology

Marine Biology

Diptanshu Roy

RANDOM PUBLICATIONS
NEW DELHI (INDIA)

Marine Biology

ISBN 978-93-5111-423-9

Published in 2014 in India by

RANDOM PUBLICATIONS

4376-A/4B, Gali Murari Lal, Ansari Road
New Delhi-110 002
Phone : +91-11-43580356, +91-11-23289044
e-mail: randomexports@gmail.com, sales@randompublications.com, info@randompublications.com

Reprinted 2023

Type Setting by: Friends Media, Delhi-110089
Printed at : Replika Press Pvt. Ltd.

Preface

Marine biology is a branch of biology and is closely linked to oceanography. It also encompasses many ideas from ecology. Fisheries science andmarine conservation can be considered partial offshoots of marine biology (as well as environmental studies). Marine Chemistry, Physical oceanography and Atmospheric sciences are closely related to this field. Marine biology is the scientific study of organisms in the ocean or other marine or brackish bodies of water. Given that in biology many phyla, families and genera have some species that live in the sea and others that live on land, marine biology classifies species based on the environment rather than on taxonomy. Marine biology differs from marine ecology as marine ecology is focused on how organisms interact with each other and the environment, and biology is the study of the organisms themselves. A large proportion of all life on Earth exists in the ocean. Exactly how large the proportion is unknown, since many ocean species are still to be discovered. The ocean is a complex three-dimensional world covering about 71% of the Earth's surface. Because of its depth it contains about 300 times the habitable volume of the terrestrial habitats on Earth. The habitats studied in marine biology include everything from the tiny layers of surface water in which organisms and abiotic items may be trapped in surface tension between the ocean and atmosphere, to the depths of the oceanic trenches, sometimes 10,000 meters or more beneath the surface of the ocean. Specific habitats include coral reefs, kelp forests, seagrass meadows, the surrounds of seamounts and thermal vents, tidepools, muddy, sandy and rocky bottoms, and the open ocean (pelagic) zone, where solid objects are rare and the surface of the water is the only visible boundary. The organisms studied range from microscopic phytoplankton and zooplankton to huge cetaceans (whales) 30 meters (98 feet) in length.

Marine life is a vast resource, providing food, medicine, and raw materials, in addition to helping to support recreation and tourism all over the world. At a fundamental level, marine life helps determine the very nature of our planet. Marine organisms contribute significantly to the oxygen cycle, and are involved in the regulation of the Earth's climate. Shorelines are in part shaped and protected by marine life, and some marine organisms even help create new land. Many species are economically important to humans, including food fish (both finfish and shellfish). It is also becoming understood that the well-being of marine organisms and other organisms are linked in very fundamental ways. The human body of knowledge regarding the relationship between life in the sea and important cycles is rapidly growing, with new discoveries being made nearly every day. These cycles include those of matter (such as the carbon cycle) and of air (such as Earth's respiration, and movement of energy through ecosystems including the ocean). Large areas beneath the ocean surface still remain effectively unexplored.

For every student, teacher and researcher in the subject it offers a solid basis for an in-depth understanding of the entire subject area.

I thank all members of my team who have helped in the preparation of the book. My special thanks go to "Random Publication" who have published the book.

—Diptanshu Roy

Contents

1

Nature of Marine Life

Biological Expeditions

Although a number of 16th and 17th century travellers provided much valuable information about the plants and animals in the Orient, America, and Africa, most of this information was collected by curious individuals rather than trained observers. A development that occurred during the 18th and 19th centuries was the organisation of scientific expeditions, usually under the auspices of a particular government. The most notable of these efforts were the voyages of the "Endeavour," the "Investigator," the "Beagle," and the "Challenger," all sponsored by the English government. Captain James Cook sailed the "Endeavour" to the South Sea islands, New Zealand, New Guinea, and Australia in 1768; the voyage provided Joseph Banks, a young naturalist, with the opportunity to make a very extensive collection of plants and notes, which helped establish him as a leading biologist. Another expedition to the same area in the "Investigator" in 1801 included a botanist, Robert Brown, whose work on the plants of Australia and New Zealand became a classic; especially important were his descriptions of how certain plants adapt to different environmental conditions. Brown is also credited with discovering the cell nucleus and analysing sexual processes in higher plants.

One of the most famous biological expeditions of all time was that of the "Beagle" in 1831, the members including Charles Darwin. Although Darwin's primary interest at the time was geology, his visit to the Galápagos Islands aroused his interest in biology and caused him to speculate about their curious insular animal life and the significance of isolation in space and time for the formation of species. During the "Beagle" voyage, Darwin collected specimens of and accumulated copious notes on the plants and animals of South America and Australia, for which he received great acclaim on his return to England. In spite of these expeditions, the contributions made by individuals were still very important. Such an individual was the English naturalist Alfred Russel Wallace, who undertook explorations of the Malay Peninsula from 1854 to 1862. In 1876 he published his book *The Geographical Distribution of Animals*, in which he divided the landmasses into six zoogeographical regions and described their characteristic fauna. Wallace also contributed to the theory of evolution, publishing in 1870 a book expressing his views, *Contributions to the Theory of Natural Selection.*

The Development of the Cell Theory

Although the microscopists of the 17th century had made detailed descriptions of plant and animal structure and though Hooke had coined the term cell for the compartments he had observed in cork tissue, their observations lacked an underlying theoretical unity. It was not until 1838 that Matthias J. Schleiden, a German botanist interested in plant anatomy, stated, "the lower plants all consist of one cell, while the higher ones are composed of (many) individual cells." When Schleiden's friend, the German physiologist Theodour Schwann, extended the cellular theory to include animals, he thereby brought about a rapprochement between botany and zoology. The formation of the cell theory—all plants and animals are made up of cells—marked a great conceptual advance in biology, and it resulted in renewed attention to the living processes that go on in cells. In 1846, after several investigators had described the streaming movement of the cytoplasm in plant cells, Hugo von Mohl, a German botanist, coined the word protoplasm to designate the living substance of the cell. The concept of protoplasm as the physical basis of life led to the development of cell physiology. A further extension of the cell theory was the development of cellular pathology by Rudolf Virchow, who established the relationship between abnormal events in the body and unusual cellular activities. This gave a new direction to the study of pathology and resulted in advances in medicine.

The Theory of Evolution

As knowledge of plant and animal forms accumulated during the 16th, 17th, and 18th centuries, a few biologists began to speculate about the ancestry of these organisms, though the prevailing view was that promulgated by Linnaeus namely, the immutability of the species. Among the early speculations voiced during the 18th century, Erasmus

Darwin, an English physician and the grandfather of Charles Darwin, concluded that species descend from common ancestors and that there is a struggle for existence among animals. A French naturalist, Jean-Baptiste Lamarck, who was probably the most important of the 18th century evolutionists, recognised the role of isolation in species formation; he also saw the unity in nature and conceived the idea of the evolutionary tree.

A complete theory of evolution was not announced, however, until the publication in 1859 of Charles Darwin's In his book Darwin stated that all living creatures multiply so rapidly that, if left unchecked, they would soon overpopulate the world. According to Darwin, the checks on population size are maintained by competition for the means of life. Hence, if any member of a species differs in some way that makes it better fitted to survive, then it will have an advantage that its offspring would be likely to perpetuate. Darwin's work reflects the influence of a British economist, Thomas Robert Malthus, who in 1838 published an essay on population in which he warned that if man multiplies more rapidly than his food supply, competition for existence would result.

Darwin was also influenced by a British geologist, Charles Lyell, who realised from his studies of geological formations that the relative ages of deposits could be estimated by means of the proportion of living and extinct mollusks. But it was not until after his travels in the "Beagle" in 1831, during which he observed a great richness and diversity of island fauna, that Darwin began to develop his theory of evolution. Alfred Russel Wallace had reached conclusions similar to those of Darwin following his studies of plants and animals in the Malay Peninsula.

Conceptually, the theory was of the utmost significance, accounting as it did for the formation of new species. Following the subsequent discovery of the chromosomal basis of inheritance and the laws of heredity, it could be seen that natural selection does not involve the sharp alternatives of life or death but results from the differential survival of variants. Today, the universal principle of natural selection, which is the central concept of Darwin's theory, is firmly established.

The Study of the Reproduction and Development of Organisms

Preformation versus Epigenesis

A question posed by Aristotle was whether the embryo is preformed and therefore only enlarges during development or whether it differentiates from an amorphous beginning. Two conflicting schools of thought had been based on this question: the preformation school maintained that the egg contains a miniature individual that develops into the adult stage in the proper environment; the epigenesis school believed that the egg is initially undifferentiated and that development occurs as a series of steps.

Prominent supporters of the preformation doctrine, which was widely held until the 18th century, included Malpighi, Swammerdam, and Leeuwenhoek. In the 19th century, as criticism of preformation mounted, Karl Ernst von Baer, an Estonian embryologist, provided the final evidence against the theory. His discovery of the mammalian egg and his recognition of the formation of the germ layers out of which the embryonic organs develop laid the foundations of modern embryology.

The Fertilization Process

Despite the many early descriptions of spermatozoa, their essential role in fertilization was not proven until 1879, when Hermann Fol, a Swiss physician and zoologist, observed the penetration of a spermatozoon into an ovum. Prior to this discovery, during the period from 1823 to 1830, the existence of the sexual process in flowering plants had been demonstrated by Giovanni Battista Amici, an Italian astronomer and botanist, and confirmed by others. The discovery of fertilization in plants was of great importance to the development of plant hybrids, which are produced by cross-pollination between different species; it was also of great significance to the studies of genetics and evolution.

The universal occurrence and remarkable similarity of the fertilization process, regardless of the organism in which it occurs, provoked many of the leading investigators of the time to search for the underlying mechanism. It was realised that there must be some way by which the number of chromosomes is reduced before fertilization; otherwise the spermatozoon fused with an egg.

In 1883 Edouard van Beneden, a Belgian cytologist, showed that the eggs and spermatozoa in the worm *Ascaris* contain half the number of chromosomes found in the body cells. To account for the halving of the chromosomes in the sex cells, a process that is called meiosis, in 1887 August Weismann, a German biologist, suggested that there must be two different types of cell division, and by 1900 the details of meiosis had been elucidated.

The Study of Heredity

Pre-Mendelian Theories of Heredity

The fundamental laws of heredity were discovered in 1865 by Gregor Mendel, an Austrian monk and biologist, but his work was ignored until its rediscovery in 1900. There were, however, a number of views on the subject that had been expressed long before Mendel. The Greek philosophers, for example, believed that the traits of individuals were acquired from contact with the environment and that such acquired characteristics could be inherited by offspring.

Because Lamarck was the most famous proponent of the inheritance of acquired characteristics, the theory is called Lamarckism. This concept, which emphasised the use and disuse of organs as the significant factor in determining the characteristics of an individual, postulated that any alterations in the individual could be transmitted to the offspring through the gametes. Yet the inheritance of acquired characteristics has never been experimentally verified, despite many attempts. Furthermore, many of Lamarck's examples, such as the long neck of the giraffe, can be more satisfactorily explained by means of natural selection.

In 1885 Weismann suggested that hereditary characteristics were transmitted by what he called germ plasm—as distinguished from the somatoplasm (body cells)—which linked the generations by a continuous stream of dividing germ cells. In stating definitely seven years later that the material of heredity was in the chromosomes, Weismann anticipated the chromosomal basis of inheritance.

Francis Galton, a 19th century English anthropologist, made a number of important contributions to genetics, one of which was a study of the hereditary nature of ability, from which he developed the concept that judicious breeding could improve the human race (eugenics). Galton's most significant work was the demonstration that each generation of ancestors makes a proportionate contribution to the total makeup of the individual. Thus, he suggested that if a tall man marries a short woman, each should contribute half of the total heritage, and the resultant offspring should be intermediate between the two parents.

Mendelian Laws of Heredity

Tall versus short; round seed versus wrinkled seed. When Mendel fertilised short plants with pollen from tall plants, he found the offspring (first filial generation) to be uniformly tall. But if he allowed the plants of this generation to self-pollinate (fertilize themselves), their offspring (the second filial generation) exhibited the characters of the grandparents in a rather consistent ratio of three tall to one short. Furthermore, if allowed to self-pollinate, the short plants always bred true—*i.e.,* never produced anything but short plants. From these results Mendel developed the concept of dominance, based on the supposition that each plant carried two trait units, one of which dominated the other. Nothing was known at that time about chromosomes or meiosis, yet Mendel deduced from his results that the trait units, later called genes, could be a kind of physical particle that was transmitted from one generation to another through the reproductive mechanism.

Mendel's most important concept was the idea that the paired genes present in the parent separate or segregate during the formation of the gametes. Moreover, in later experiments in which he studied the inheritance of two pairs of traits, Mendel showed that one pair of genes is independent of another. Thus, the principles of segregation and of independent assortment were established.

Mendel's findings were ignored for 35 years, probably for two reasons. Because the distinguished Swiss botanist Karl Wilhelm von Nägeli failed to recognise the significance of the work after Mendel had sent him the results, he did nothing to encourage Mendel. Nägeli's great prestige and the lack of his endorsement indirectly weighed against widespread recognition of Mendel's work. Moreover, when the work was published, little was known about the cell, and the processes of mitosis and meiosis were completely unknown. Mendel's work was finally rediscovered in 1900, when three botanists independently recognised the worth of his studies from their own research and cited his publication in their work.

Elucidation of the Hereditary Mechanism

By 1901 it was understood how the hereditary units postulated by Mendel are distributed; it was also known that the somatic (body) cells have a double, or diploid, complement of chromosomes, while the reproductive cells have a single, or haploid, chromosome number suggested that each chromosome in a pair can exchange the hereditary factors it carries with those of the other chromosome. At first the U.S. geneticist Thomas Hunt Morgan dismissed this concept, but later, when he found that it agreed with his own laboratory findings, Morgan and his collaborators assigned the hereditary units (genes) specific positions, or loci, within the chromosomes. With the genes established as the carriers of hereditary traits, William Bateson, an English biologist, coined the name genetics for the experimental study of heredity and evolution.

Biology in the 20th Century

Just as the 19th century can be considered the age of cellular biology, the 20th century has been characterised by developments in molecular biology.

Important Conceptual Developments

By utilising modern methods of investigation, such as X-ray diffraction and electron microscopy, to explore levels of cellular organisation beyond that visible with a light microscope—*i.e.,* the ultrastructure of the cell—new concepts of cellular function have been produced. Not only has the study of the molecular organisation of the cell probably had the greatest impact upon biology during the 20th century but it also has led directly to the convergence of many different scientific disciplines in order to acquire a better understanding of life processes.

Another 20th century development has been the realisation that man is as dependent upon the Earth's natural resources as are other animals. The progressive destruction of the environment can be attributed, in part, to an increase in population pressure as well as to certain technological advances. Thus, though lifesaving advances in medicine have resulted in a dramatic drop in the death rate, they have also been a factor contributing to the explosive increase in the human population. Moreover, chemical contaminants being introduced into the environment by manufacturing processes, pesticides, automobile emissions, and other means are seriously endangering all forms of life. It is for these reasons that biologists are beginning to pay much greater attention to the relationships of living things to each other as well as to their biotic and abiotic environments.

Intradisciplinary Work

There are many important categories in the biological sciences. Botany, zoology, and microbiology deal with types of organisms and their relationships with each other. Such disciplines are subdivided into more specialised categories; for example, ichthyology is the study of fishes, algology the study of algae. All of them draw upon paleontology, taxonomy, morphology, and evolution. In the past few decades, many developments in physiology and embryology have resulted from studies in cell biology, biophysics, and biochemistry.

This has given rise to cell physiology, cytochemistry, and ultrastructural studies, which aim at correlating structure with function. Ecology, the study of the relations of a group of organisms to its environment, includes both the physical features of the environment and other organisms that may compete for food and shelter. Ecology may be subdivided according to the environment—for example, freshwater ecology and marine ecology—and draws upon animal behaviour. One aspect of cell biology, formerly called cytology, is the investigation of the structure, composition, and function of cells; biochemistry and biophysics provide important information.

Thus, biology encompasses a number of disciplines; in fact, it has become common to divide biology into its several levels of organisation rather than separating the disciplines. It is useful, for example, to differentiate between organismic biology, the study of the whole organism, and cell biology. Similarly the technological advances of the 20th century have allowed increased understanding of the molecules comprising living things and their aggregation and organisation into such structures as chromosomes and membranes. Knowledge of this aspect, called molecular biology, represents the molecular level of organisation. The fourth level, population biology, involves the complex interaction of population of animals and plants with the environment.

Relations with Other Disciplines

In the 17th century, with the invention of the microscope, which made possible study of the cellular level of organisation, biology began to receive the benefits of scientific developments in physics. In the 18th century such developments in chemistry as a better understanding of the nature of oxygen, carbon dioxide, and water began to have important implications for biology. Today, through the disciplines of biochemistry and biophysics, both chemistry and physics have continued to make significant contributions to biology, particularly in the area of molecular biology.

Biology is also very closely related to the disciplines of medicine and agriculture, out of which it developed as an independent discipline. In a sense, the roles have been reversed in the 20th century, for it is basic research being conducted in biology that is contributing to major advances currently being made in medicine and agriculture. It was biological

Another scientific discipline, that of geology, is closely related to the biological study of paleontology. The technique of radiocarbon dating, which was developed by chemists to determine the age of biological remains, has been of great use in the fields of archaeology and anthropology as well as biology. A new discipline, space biology, has arisen through the activities of the scientists and engineers concerned with the exploration of space.

The conceptual framework of biology has had to be altered to accommodate newly discovered facts. In the process biology has received contributions from and made contributions to many other disciplines, in the humanities as well as in the sciences.

Changing Social and Scientific Values

The biologist's role in society as well as his moral and ethical responsibility in the discovery and development of new ideas has led to a reassessment of his social and scientific value systems. A scientist can no longer ignore theconsequences of his discoveries; he is as concerned with the possible misuses of his findings as he is with the basic research in which he is involved.

This emerging social and political role of the biologist and all other scientists requires a weighing of values that cannot be done with the accuracy or the objectivity of a laboratory balance. As a member of society, it is necessary for a biologist now to redefine his social obligations and his functions, particularly in the realm of making judgments about such ethical problems as man's control of his environment or his manipulation of genes to direct further evolutionary development.

The Impact of Microbes on the Environment and Human Activities

Beneficial Effects of Microorganisms

Microbes are everywhere in the biosphere, and their presence invariably affects the environment that they are growing in. The effects of microorganisms on their environment can be beneficial or harmful or inapparent with regard to human measure or observation.

Since a good part of this text concerns harmful activities of microbes (i.e., agents of disease) this chapter counters with a discussion of the beneficial activities and exploitations of microorganisms as they relate to human culture. The beneficial effects of microbes derive from their metabolic activities in the environment, their associations with plants and animals, and from their use in food production and biotechnological processes.

Nutrient Cycling and the Cycles of Elements that Make Up Living Systems

At an elemental level, the substances that make up living material consist of carbon (C), hydrogen (H), oxygen (O), nitrogen (N), sulfur (S), phosphorus (P), potassium (K), iron (Fe), sodium (Na), calcium (Ca) and magnesium (Mg). The primary constituents of organic material are C, H, O, N, S, and P. An organic compound always contains C and H and is symbolised as CH_2O (the empirical formula for glucose). Carbon dioxide (CO_2) is considered an inorganic form of carbon.

The most significant effect of the microorganisms on earth is their ability to recycle the primary elements that make up all living systems, especially carbon (C), oxygen (O) and nitrogen (N). These elements occur in different molecular forms that must be shared among all types of life. Different forms of carbon and nitrogen are needed as nutrients by different types of organisms. The diversity of metabolism that exists in the microbes ensures that these elements will be available in their proper form for every type of life. The most important aspects of microbial metabolism that are involved in the cycles of nutrients are discussed below.

Primary production involves photosynthetic organisms which take up CO_2 in the atmosphere and convert it to organic (cellular) material. The process is also called CO_2 fixation, and it accounts for a very large portion of organic carbon available for synthesis of cell material. Although terrestrial plants are obviously primary producers, planktonic algae and cyanobacteria account for nearly half of the primary production on the planet. These unicellular organisms which float in the ocean are the "grass of the sea", and they are the source of carbon from which marine life is derived.

NASA receives data from the Terra and Aqua satellites which measures net primary productivity on Earth. These false colour maps represents the rate at which photosynthetic organisms absorb carbon out of the atmosphere. The yellow and red areas show the highest rates, ranging from 2 to 3 kilograms of carbon taken in per square metre per year. The green, blue, and purple shades show progressively lower productivity. Tropical rain forests are generally the most productive places on Earth. However, primary productivity near the seaï¿½s surface over such a widespread area of the Earth makes the ocean roughly as productive as the land.

Decomposition or biodegradation results in the breakdown of complex organic materials to forms of carbon that can be used by other organisms. There is no naturally-occurring organic compound that cannot me degraded by some microbe, although some synthetic compounds such as teflon, styrofoam, plastics, insecticides and pesticides are broken down slowly or not at all. Through the metabolic processes of fermentation and respiration, organic molecules are eventually broken down to CO_2 which is returned to the atmosphere.

Waste management, whether in compost, landfills or sewage treatment facilities, exploits activities of microbes in the carbon cycle. Organic (solid) materials are digested by microbial enzymes into substrates that eventually are converted to a few organic acids and carbon dioxide.

Nitrogen fixation is a process found only in some bacteria which removes N_2 from the atmosphere and converts it to ammonia (NH_3), for use by plants and animals. Nitrogen fixation also results in replenishment of soil nitrogen removed by agricultural processes. Some bacteria fix nitrogen in symbiotic associations in plants. Other Nitrogen-fixing bacteria are free-living in soil and aquatic habitats

Some habitats like this cactus community in the Sonoran Desert, rely on nitrogen-fixing bacteria at the base of the food chain as the source of nitrogen for maintenance of cell material. Every plant in

Oxygenic photosynthesis occurs in plants, algae and cyanobacteria. It is the type of photosynthesis that results in the production of O_2 in the atmosphere. At least 50 percent of the O_2 on earth is produced by photosynthetic microorganisms (algae and cyanobacteria), and for at least a billion years before plants evolved, microbes were the only organisms producing O_2 on earth. O_2 is required by many types of organisms, including animals, in their respiratory processes.

The cyanobacterium, *Synechococcus*, is a primary component of marine and freshwater plankton and microbial mats, The unicellular procaryote is involved in primary production, nitrogen fixation and oxygenic photosynthesis and thereby participates in the cycles of carbon, nitrogen and oxygen. *Synechococcus* is among the most important photosynthetic bacteria in marine environments, estimated to account for about 25 percent of the primary production that occurs in typical marine habitats.

Biogeography, Ecology, and Vulnerability of Chemosynthetic Ecosystems in the Deep Sea

This chapter is based upon research and findings relating to the Census ofMarine Life ChEss project, which addresses the biogeography of deep-water chemosynthetically driven ecosystems This project has been motivated largely by scientific questions concerning phylogeographic relationships among different chemosynthetic habitats, evidence of conduits and barriers to gene flow among those habitats, and environmental factors that control diversity and distribution of chemosynthetically driven fauna. Investigations of chemosynthetic environments in the deep sea span just three decades, owing to their relatively recent discovery.

Despite the excitement of many discoveries in the deep ocean since the early nineteenth century, nothing could have prepared the scientific community for the discovery made in the late 1970s, which would challenge some fundamental principles of our understanding of life on Earth. Deep hot water venting was observed for the first time in 1977 on the Galápagos Rift, in the eastern Pacific. To the astonishment of the deep-sea explorers of the time, a prolific community of bizarre animals were seen to be living in close proximity to these vents (Corliss et al. 1979). Giant tubeworms and huge white clams were among the inhabitants, forming oases of life in the otherwise apparently uninhabited deep seafloor (Figs. 9.1A, B, and C). Most of the creatures first observed on vents were totally new to science, and it was a complete mystery as to what these animals were using for an energy source in the absence of sunlight and in the presence of toxic levels of hydrogen sulfide and heavy metals.

Chemosynthetic Ecosystems: Where Energy from the Deep Seabed is the Source of Life

Until the discovery of hydrothermal vents, benthic deep-sea ecosystems were assumed to be entirely heterotrophic, completely dependent on the input of sedimented organic matter produced in the euphotic surface layers from photosynthesis (Gage 2003) and, in the absence of sunlight, completely devoid of any in situ primary productivity. The deep sea is, in general, a food-poor environment with low secondary productivity and biomass. In 1890, Sergei Nikolaevich Vinogradskii proposed a novel life process called chemosynthesis, which showed that some microbes have the ability to live solely on inorganic chemicals. Almost 90 years later the discovery of hydrothermal vents provided stunning new insight into the extent to which microbial primary productivity by chemosynthesis can maintain biomass-rich metazoan communities with complex trophic structure in an otherwise food-poor deep sea (Jannasch & Mottl 1985). Hydrothermal vents are found on mid-ocean ridges and in back arc basins where deep-water volcanic chains form new ocean floor (reviewed by Van Dover 2000; Tunnicliffe et al. 2003).

The super-heated fluid (up to 407 °C) emanating from vents is charged with metals and sulfur. Microbes in these habitats obtain energy from the oxidation of hydrogen, hydrogen sulfide, or methane from the vent fluid. The microbes can be found either suspended in the water column or forming mats on different substrata, populating seafloor sediments and ocean crust, or living in symbiosis with several major animal taxa (Dubilier et al. 2008; Petersen & Dubilier 2009). By microbial mediation, the rich source of chemical energy supplied from the deep ocean interior through vents allows the development of densely populated ecosystems, where abundances and biomass of fauna are much greater than on the surrounding deep-sea floor.

Eight years after the discovery of hydrothermal vent communities, the first cold seep communities were described in the Gulf of Mexico (Paull et al. 1984). Cold seeps occur in both passive and active (subduction) margins. Seep habitats are characerised by upward flux of cold fluids enriched in methane and often also other hydrocarbons, as well as a high concentration of sulfide in the sediments (Sibuet & Olu 1998; Levin 2005). The first observations of seep communities showed a fauna and trophic ecology similar to that of hydrothermal vents at higher taxonomical levels (Figs. 9.1D and E), but with dissimilarities in terms of species and community structure.

The energetic input to chemosynthetic ecosystems in the deep sea can also derive from photosynthesis as in the case of large organic falls to the seafloor, including kelp, wood, large fish, or whales. After a serendipitous discovery of a whale fall in 1989, the first links between vents, seeps, and the reducing ecosystems at large organic falls were made (Smith & Baco 2003). Bones of whales consist of up to 60% lipids that, when degraded by microbes, produce reduced chemical compounds similar to those emanating from vents and seeps(Treude et al. 2009). Another deep-water reducing environment is created where oxygen minimum zones (OMZs, with oxygen concentrations below 0.5 ml l^{-1} or 22 μM) intercept continental margins, occurring mainly beneath regions of intensive upwelling (Helly & Levin 2004) (Figs. 9.1G and H). Only in the second half of the twentieth century was it understood that OMZs support extensive autotrophic bacterial mats (Gallardo 1963, 1977; Sanders 1969; Fossing et al. 1995; Gallardo & Espinoza 2007) and, in some instances, fauna with a trophic ecology similar to that of vents and seeps (reviewed in Levin 2003).

Adaptations to an "Extreme" Environment

Steep gradients of temperature and chemistry combined with a high disturbance regime, caused by waxing and waning of fluid flow and other processes during the life cycle of a hydrothermal vent, result in low diversity communities with only a few mega- and macrofauna species dominating any given habitat (Van Dover & Trask 2001; Turnipseed et al. 2003; Dreyer et al. 2005). The proportion of extremely rare species (fewer than five individuals in pooled samples containing tens of thousands of individuals from the same vent habitat) is typically high, in the order of 50% of the entire species list for a given quantitative sampling effort (C.L. Van Dover, unpublished observation).

Deep-water chemosynthetic habitats have also been shown to have a high degree of species endemicity in each habitat: 70% in vents (Tunnicliffe et al. 1998; Desbruyères et al. 2006a), about 40% in seeps both for mega epifauna (Bergquist et al. 2005; Cordes et al. 2006) and macro infauna (Levin et al. 2009a). In OMZs, the percentage of endemism is relatively low (Levin et al. 2009a), but has yet to be quantified. Some of the most conspicuous of the endemic species of reducing environments have developed unusual physiological adaptations for the extreme environments in which they live.

These include symbiotic relationships with bacteria, organ and body modifications, and reproductive and novel adaptations for tolerating thermal and chemical fluctuations of great magnitude. Because chemosynthetic habitats are naturally fragmented and ephemeral habitats, successful species must also be specially adapted for dispersal to and colonisation of isolated "chemosynthetic islands" in the deep sea (Bergquist et al. 2003; Neubert et al. 2006; Vrijenhoek 2009a).

Chemosynthetic Islands: A Biogeographic Puzzle with Missing Pieces

Since their discovery just over 30 years ago, more than 700 species from vents (Desbruyères et al. 2006a) and 600 species from seeps have now been described and are listed on ChEssBase (Ramirez-Llodra et al. 2004; This rate of discovery is equivalent to one new species described every two weeks, sustained over approximately one-quarter of the past century (Lutz 2000; Van Dover et al. 2002). Furthermore, geomicrobiologists have explored microbial diversity of chemosynthetic ecosystems, revealing a plethora of interesting and novel metabolisms, but also signature compositions for the different types of reduced habitat, and symbiotic organism (Jørgensen & Boetius 2007; Dubilier et al. 2008).

Although several hundred hydrothermal vent and cold seep sites have now been located worldwide, only approximately 100 have been studied so far with respect to their faunal and microbial composition, and even for their ecosystem function. Nevertheless, through such investigations, scientists soon noticed the differenes and in some cases similarities among the animal communities from different vent and seep sites.

For example, why is the giant tubeworm *Riftia pachyptila* only found at Pacific vents whereas shrimp species in the genus *Rimicaris* are only found at Atlantic and Indian Ocean vents? Why is the mussel genus *Bathymodiolus* generally widespread at vents and seeps but largely absent from seeps and vents in the northeastern Pacific Ocean? In 2002, at the onset of the ChEss project, biological investigations of known vent sites provided enough data to describe six biogeographic provinces for vent species (Van Dover et al. 2002) and identified several gaps that needed to be closed to complete the "biogeographical puzzle of seafloor life" (Shank 2004). In contrast, cold seep and whale fall communities appear to share many of the key taxa across all oceans. The ChEss project developed a major exploratory program to address and explain global patterns of biogeography in deep-water chemosynthetic ecosystems and the factors shaping them.

Technological Developments for Exploration

One of the most significant advances in deep-sea investigations of chemosynthetic ecosystems, developed and implemented as a new international state of the art technique within the lifetime of the ChEss project, has been the use of deep-sea autonomous underwater vehicles (AUVs) to trace seafloor hydrothermal systems to their source or to map cold seep systems in the necessary resolution to quantify the distribution of chemosynthetic habitats. This approach (Baker et al. 1995; Baker & German 2004; Yoerger et al. 2007) was sufficient for geological investigations of global-scale heat-flux and chemical discharge to the oceans. However, the ChEss hypotheses concerning global-scale biogeography required more precise location of hydrothermal venting and hydrocarbon seepage on the seafloor; ideally with preliminary characterisation of not only the vent and seep site itself but also a first-order characterisation of the dominant species present. So far, the method has been applied on seven separate hydrothermal vent cruises, from 2002 to 2009, throughout the Southern hemisphere, the least explored part of the global deep ocean. These expeditions have located 16 different new sites on the Galápagos Rift (Shank et al. 2003), in the Lau Basin (southwest Pacific; German et al. 2008a), th Mid-Atlantic Ridge (MAR) (South Atlantic; German et al. 2008b; Melchert et al. 2008; Haase et al. 2009), the southwest Indian Ridge (Southern Indian Ocean; C. Tao, personal communication), the East Pacific Rise (southeast Pacific; C. Tao, personal communication), and the Chile margin (C. German, unpublished observation). For cold seep mapping, a major success was the combined AUV and remotely operated vehicle (ROV) deployment in the Nile Deep Sea Fan, leading to the description of several new types of hydrocarbon seep in depths between 1,000 and 3,500 m (Foucher et al. 2009; technical details described in Dupré et al. 2009).

The way the AUV technique works for the exploration of vents is described in detail by German et al. (2008a). Perhaps most surprising to us, and of widest long-term significance, is that, when flying close to the seafloor, the techniques have not only been sufficiently sensitive to locate high-temperature "black-smoker" venting, but also sites of much more subtle lower-temperature diffuse flow (Shank et al. 2003). Building on these successes, future investigations will be reliant upon the new generation of exploratory vehicles such as a new hybrid AUV–ROV vehicle (Bowen et al. 2009), which has already been applied in ChEss studies as a technological precursor to future under ice investigations (Jakuba et al. 2008; German et al. 2009).

Finding New Species

In the past decade, we have seen a significant increase in molecular tools for studies to understand species evolution, metapopulations, and gene flow in chemosynthetic regions (Shank & Halanych 2007; Johnson et al. 2008; Plouviez et al. 2009; Vrijenhoek 2009b). New high-resolution and high-throughput methods will result in the first insight into the structure and biogeography of microbial communities of chemosynthetic ecosystems in the Census International Census f Marine Microbes (ICoMM) project. However, a major concern today for marine biodiversity analysis is the paucity of taxonomists using morphological methods, and in particular taxonomists specialising in deep-sea species. Both morphological and molecular taxonomy are essential to develop fundamental knowledge and sustainable management of our marine resources. In an effort to raise the profile of taxonomy once more, ChEss set up an annual program of Training Awards for New Investigators (TAWNI).

These awards have been made to a total of 10 scientists from around the globe to develop further their taxonomic skills relating to chemosynthetic organisms). As a result, they have collectively achieved impressive outputs where many meio-, macro-, and megafauna species have been described and new records identified from different sites. These descriptions have been added to the approximately 200 species that have been described and published from vents, seeps, and whale falls since the onset of the ChEss project in 2002. One of the most extraordinary animals that has consequently received much media attention was discovered on southeast Pacific vents in 2005: the yeti crab *Kiwa hirsuta*. This is not only a species new to science, but also represents a new genus and new family (Macpherson et al. 2005). Recently, a close relative of the vent yeti crab was discovered from Costa Rican cold seeps and is being described with the aid of a TAWNI grant (A. Thurber, personal communication).

Global Biogeography Patterns in Deep-Water Chemosynthetic Ecosystems

Addressing global biogeographic patterns for species from all deep-water chemosynthetic ecosystems and the phylogenetic links among habitats needed a coordinated international effort, with shared human and infrastructure resources, that no single nation could attempt alone. In 2002, ChEss outlined a field program for the strategic exploration and investigation of chemosynthetic ecosystems in key areas that would provide essential information to close some of the main gaps in our knowledge (Tyler et al. 2003).

The ChEss field program was motivated by three scientific questions. (1) What are the taxonomic relationships among different chemosynthetic habitats? (2) What are the conduits and barriers to gene flow among those habitats? (3) What are the environmental factors that control diversity and distribution of chemosynthetically driven fauna? To address these questions at the global scale, four key geographic areas were selected for exploration and investigation: the Atlantic Equatorial Belt (AEB), the New Zealand Region (RENEWZ), the Polar Regions (Arctic and Antarctic), and the southeast Pacific off Chile region (INSPIRE). Below, we describe the issues addressed and main findings in each area.

The Atlantic Equatorial Belt: Barriers and Conduits for Gene Flow

The AEB is a large region expanding from Costa Rica to the West Coast of Africa that encloses numerous seep (for example Costa Rica,

Gulf of Mexico, Blake Ridge, Gulf of Guinea) and vent (e.g., northern MAR (NMAR), southern MAR (SMAR), Cayman Rise) habitats. This region is particularly significant for investigating connectivity among populations and species' maintenance across large geographic areas. Potential gene flow across the Atlantic (west to east) is subject to the effects of deep-water currents (Northeast Atlantic deep water), equatorial jets, and topographic barriers such as the MAR. When considering a north–south direction, gene flow along the MAR may be affected by mid-ocean ridge offsets such as the Romanche and Chain fracture zones.

These fracture zones are significant topographic features 60 million years old, 4 km high and 935 km ridge offset, which cross the equatorial MAR prominently, affecting both the linearity of the ridge system and large-scale ocean circulation in this region. North Atlantic Deep Water flows south along the East coasts of North and South America as far as the Equator before being deflected east, crossing the MAR through conduits created by these major fracture zones (Speer et al. 2003). Circulation within these fracture zones is turbulent and may provide an important dispersal pathway for species from west to east across the Atlantic (Van Dover et al. 2002), for example between the Gulf of Mexico and the Gulf of Guinea.

The cold seeps in the Pacific Costa Rican margin were included in this study to address questions of isolation between the Pacific and the Atlantic faunas after the closure of the Isthmus of Panama 5 million years ago. The fauna from methane seeps on the Costa Rica margin, just now being explored, are yielding surprising affinities, which suggests that this site operates as a crossroads. Some animals appear related to the seep faunas in the Gulf of Mexico and off West Africa,whereas others show phylogenetic affinities with nearby vents at 9° N on the East Pacific Rise and with more distant vents at Juan de Fuca Ridge and the Galápagos (L. Levin, unpublished observations). Furthermore, recent investigations have shown (C. German, C.L. Van Dover & J. Copley, unpublished observation) there is active venting in the ultra-slow Cayman spreading ridge in the Caribbean at depths of 5,000 m (CAYTROUGH 1979), and investigations are underway to determine how the animals colonising these vents are related to vent and seep faunas on either side of the Isthmus of Panama. The first plumes were located in November 2009 at depths below 4,500 m, suggestive of active venting, and these plumes were further explored by ChEss scientists in 2010 who located the source of active venting at 5,000m – the deepest known vent ever found. Exploration on the

MAR has also led to the discovery of the hottest vents (407 °C) (Kochinsky 2006; Kochinsky et al. 2008), as well as another deep vent (4,100 m), named Ashadze (Ondreas et al. 2007; Fouquet et al. 2008).

In the AEB, the connections across the Atlantic have been relatively clearly defined. The seeps of the African margin contain a fauna with very close affinities to the seep communities in the Gulf of Mexico (Cordes et al. 2007; Olu-Le Roy et al. 2007; Warén & Bouchet 2009) and the seep communities on the Blake Ridge and Barbados accretionary wedge.

The communities are dominated by vestimentiferan tubeworms and bathymodioline mussels and the common seep-associated families of galatheid crabs and alvinocarid shrimp. The bathymodioline species complexes on both sides of the Atlantic sort out among the same species groupings within the genus *Bathymodiolus*, with *B. heckerae* from the Gulf of Mexico and Blake Ridge and *Bathymodiolus* sp. 1 from the African Margin in one grouping, and *B. childressi* from the Gulf of Mexico and *Bathymodiolus* sp. 2 from Africa in another group (Cordes et al. 2007). However, our understanding of the biogeographic puzzle beyond this is less clear and requires further investigation (E. Cordes, unpublished observation).

The discovery of vent sites on the southern MAR (Haase et al. 2007, 2009; German et al. 2008b) and the morphological similarity of their fauna to that of NMAR vents, suggest that the Chain and Romanche fracture zones are less of an impediment to larval dispersal than previously hypothesized (Shank 2006; Haase et al. 2007). Further support for unhindered dispersal of vent fauna along the MAR comes from molecular analyses of the two dominant invertebrates of MAR vents, *Rimicaris* shrimp and *Bathymodiolus* mussels. These studies showed recent gene flow across the equatorial zone for these key host species and their symbionts (Petersen & Dubilier 2009; Petersen et al. 2010). In addition to finding known species, new species, including the shrimp *Opaepele susannae* (Komai et al. 2007), have been described, and now more than 17 (morpho-) species have been identified from the SMAR. Many of these have been genetically compared with taxonomically similar fauna on the NMAR and reveal significant genetic divergence among species considered "the same" in both regons (T. Shank, unpublished observation)

New Zealand Region: Phylogenetic Links Among Habitats

The New Zealand region hosts a wide variety of chemosynthetic ecosystems, all in close geographic proximity. During the ChEss/

COMARGE New Zealand field program (RENEWZ), more than 10 new seep sites were discovered off the New Zealand North Island (Baco-Taylor et al. 2009). One of these sites (Builder's Pencil) covers 135,000 m^2, making it one of the largest known seep sites in the world. These initial and ongoing research activities aim at locating the sites, describing their environmental characteristics, investigating their fauna, and determining potential phylogeographic relationships among species from vents, seeps, and whale falls found in close proximity to one another.

In the New Zealand region, sampling and description of chemosynthetic communities is in its infancy. Baco-Taylor et al. (2009) have now provided an initial characterisation of cold seep faunal communities of the New Zealand region.

Preliminary biological results indicate that, although at higher taxonomic levels (family and above) faunal composition of vent and seep assemblages in the New Zealand region is similar to that of other regions, at the species level, several taxa are apparently endemic to the region. Bathymodiolin mussels and an eolepadid barnacle dominate (in number and biomass) at vent sites on the seamounts of the Kermadec volcanic arc. Genetic analysis of mussels from chemosynthetic habitats by Jones et al. (2006) revealed the New Zealand vent mussel *Gigantidas gladius* to be closely related to specis from New Zealand and Atlantic cold seeps.

New Zealand vents are often characterised by the barnacle Vulcanolepus osheai (Buckeridge 2000), found at very high densities which is different from those found farther north in the Pacific, and is most similar to an undescribed species found on the Pacific–Antarctic Ridge (Southward & Jones 2003). The most abundant motile species at Kermadec vent sites are caridean shrimp, including two species of endemic alvinocarids (*Alvinocaris niwa, A. alexander*), and one hippolytid (*Lebbeus wera*) (Webber 2004; Ahyong 2009), as well as two species of alvinocarid found elsewhere in the western Pacific (*A. longirostris, Nautilocaris saintlaurentae*; Ahyong 2009).

Only two species of low abundance and sparsely distributed vestimentiferan worms have been sampled so far from Kermadec vent sites (Miura & Kojima 2006). Of these species, *Lamellibrachia juni* has been found elsewhere in the western Pacific, whereas the other species, *Oasisia fujikurai* , is closely related to *O. alvinae* from the eastern Pacific (Kojima et al. 2006). Other species of macro and megafauna found associated with Kermadec vent sites (Glover et al.

2004; Anderson 2006; Schnabel & Bruce 2006; McLay 2007; Munroe & Hashimoto 2008; Buckeridge 2009) suggest that levels of species endemism in the New Zealand region are relatively high, although some species are either closely related to species, or are found, elsewhere in the wider Pacific region. Community-level analysis (an update of the analysis of Desbruyères et al. (2006b)) suggests that although the New Zealand region does apparently contain a vent community with a distinct composition, there is a degree of similarity with communities from elsewhere in the western Pacific (A. Rowden, unpublished observation).

In total, the analysis of samples either compiled or collected as part of the ChEss/COMARGE project provide some support for the hypothesis that the region may represent a new biogeographic province for both seep and vent fauna. However, there is clearly a need for further sampling.

Exploring Remote Polar Regions

The Polar Regions have received an increasing interest in the first decade of the twenty-first century, facilitated by new AUV technologies being developed to work in these remote areas of difficult access caused by ice coverage (Shank 2004; Jakuba et al. 2008). The exploration of the Arctic Ocean revealed, in 2003, evidence for abundant hydrothermal activity on the Gakkel Ridge (Edmonds et al. 2003). The Gakkel Ridge is an ultra-slow spreading ridge, which lies beneath permanent ice cover within the bathymetrically isolated Arctic Basin. The deep Arctic water is isolated from deep-water in the Atlantic by sills between Greenland and Iceland and between Iceland and Norway. This has important implications for the evolution and ecology of the deep-water Arctic vent fauna.

In July/August 2007, the AGAVE (Arctic GAkkel Vent Exploration) project investigated the Gakkel Ridge using AUVs and a video-guided benthic sampling system (Camper). Investigations suggested "recent" and explosive volcanic activity (Sohn et al. 2008). Extensive fields and pockets of yellow microbial mats dominated the landscape. Microbial samples revealed highly diverse chemolithotrophic microbial communities fueled by iron, hydrogen, or methane (E. Helmke, personal communication). Macrofauna associated with these mats included shrimp, gastropods, and amphipods with hexactinellid sponges peripherally attached to "older lavas". These communities may be sustained by weak fluid discharge from cracks in the young volcanic surfaces (T. Shank, unpublished observation).

So far, the northernmost vent sites that have been investigated by ROV are at 71° N on the Mohns Ridge (Schander et al. 2009). The shallow (500–750 m) sites located there support extensive mats of sulfur-oxidising bacteria. However, of the 180 species described from two fields explored, the only taxon that is potentially symbiont-bearing is a small gastropod, *Rissoa* cf. *griegi*, also known from seeps and wood falls in the North Atlantic. Arctic cold seeps have also been investigated at the Haakon Mosby Mud Volcano (HMMV) on the Barents Sea slope (72° N) at 1,280 m depth (Niemann et al. 2006; Vanreusel et al. 2009). This site has large extensions of bacterial mats and is dominated by siboglinid tubeworms (Lösekann et al. 2008), with many small bivalves of the family Thyasiridae living among them. In terms of macrofauna, the HMMV is dominated by polychaetes, with higher abundances and diversity at the siboglinid fields compared to the bacterial mats.

The meiofauna is dominated by benthic copepods in the active centre, whereas the nematode *Halomonhystera disjuncta* dominates in the bacterial mats (Van Gaever et al. 2006). The other Nordic margin cold seeps of the Storegga and Nyegga systems are also characterised by a high abundance of potentially endemic siboglinid tubeworms in association with methane seepage, as well as the occurrence of diverse mats of giant sulfide-oxidising bacteria, attracting large numbers of meio and macrofauna (Vanreusel et al. 2009).

In the Southern Ocean, vent exploration in the East Scotia Arc and seep investigations on the Weddell Sea have addressed the role of the Circumpolar Current in dispersal of deep-water fauna as a conduit between the Pacific and the Atlantic, or as a barrier between these two oceans and the Southern Ocean. The ChEss (ChEss in the Southern Ocean) project explored the East Scotia Ridge in 2009 and 2010, providing further detail to the vent plume data described by German et al. (2000). A follow-up cruise is planned for 2011 to investigate further and locate the vent source and any potential vent fauna.

On the continental margin of the Antarctic Peninsula in the Weddell Sea, cold seep communities were discovered in 800 m water beneath what was the Larsen B ice shelf. The site was once covered by extensive areas of bacterial mat and beds of live vesicomyid clams (Domack et al. 2005), but hydrocarbon seepage appeared extinct only a few years later (Niemann et al. 2009). The discovery of vesicomyid clams is evidence that the hydrographic boundary between the southern Atlantic and Pacific Oceans with Southern Ocean is not a biogeographic barrier, at least for this taxon, though it remains to be determined if the Weddell Sea vesicomyid has been sufficiently isolated to be genetically distinct from any other vesicomyid species. An international team has recently returned to the Larsen B seep sites to determine the phylogeographic alliance of the Weddell Sea clams with other vesicomyids from the Atlantic and Pacific Basins.

Southeast Pacific Off Chile: A Unique Place on Earth

The southeast Pacific region off Chile is of high interest for deep-water chemosynthetic studies, especially for the study of inter-habitat connectivity through migration and colonisation. Only here can we expect to find every known form of deep-sea chemosynthetic ecosystem in very close proximity to one another. A key reason for this unique juxtaposition of chemosynthetic habitats is the underpinning plate-tectonic setting. The Chile Rise is one of only two modern sites where an active ridge crest is being swallowed by a subduction zone and the only site where such subduction is taking place beneath a continental margin (Cande et al. 1987; Bangs & Cande 1997).

Consequently, one would expect to find hydrothermal vent sites along the East Chile Rise, cold seeps associated with subduction along the Peru–Chile trench at the intersection with the Chile Rise, and an oxygen minimum zone that abuts and extends south along the Peru and Chile margins (Helly & Levin 2004). Along with these geologic/oceanographic occurrences, significant whale feeding grounds and migration routes occur on the southwest American margin (Hucke-Gaete et al. 2004), and there is strong potential for wood-fall from the forests of southern Chile as the Andes slope steeply into the ocean south off approximately 45° S (V. Gallardo, personal communication).

To what extent will the same chemosynthetic organisms be able to take advantage of the chemical energy available at all of these diverse sites? Alternatively, will each type of chemosynthetic system host divergent fauna based on additional factors (for example depth, longevity of chemically reducing conditions, extremes of temperature, and/or fluid compositions)? Some seep sites are already known further north along the margin (Sellanes et al. 2004) and first evidence for hydrothermal activity on the medium-fast spreading Chile Rise was suggested by metalliferous input to sediments in this region (Marienfeld & Marching 1992). Systematic exploration at the very intersection of the ridge-crest and adjacent margin has recently been conducted during a joint ChEss–COMARGE cruise (February–March 2010) and sources of venting were recorded, along with evidence of at least one cold seep site relatively close by, thereby confirming expectations of the scientists on board (A. Thurber, personal communication). The Peru–Chile margin and subduction zone contain hydrate deposits and seep sites venting methane-rich fluids (Brown et al. 1996; Grevemeyer et al. 2003; Sellanes et al. 2004). Until recently these habitats had only been sampled remotely by trawl. A recent expedition provided the first Chile seep images and quantitative samples using a video-guided multicorer (A. Thurber, personal communication).

Based on trawl collections, seeps of the Chilean Margin appear to have evolved in relative isolation from other chemosynthetic communities. There are at least eight species of symbiotic bivalves, including vesicomyids, thyasirids, solemyids, and lucinids, and at least one species of the tubeworm *Lamellibrachia* (Sellanes et al. 2008). The bivalve species do not appear to have close affinities to the chemosynthetic fauna of other seeps, in particular the seeps off the coast of Peru (Olu et al. 1996) or New Zealand (Baco-Taylor et al. 2009). The general composition of the community, including a high diversity of vesicomyids (four species) is similar to that of other seep sites of the eastern Pacific. Further taxonomic resolution of this key fauna along with a complete analysis of the Costa Rica fauna will help to refine further the location of this biogeographic puzzle piece.

Larval Ecology: Shaping Faunal Distribution Under Ecological Timescales

The biological communities that inhabit chemosynthetic environments face several challenges that arise from the peculiarities of the habitat. Firstly, relatively few species are specifically adapted to the physical and chemical characteristics of these habitats. Secondly, these habitats are generally ephemeral at decadal scales (with the notable exception of seeps and OMZs) either because they are geologically unstable (vents) or because they are short lived (large organic falls). Thirdly, these habitats are patchy and can be separated by hundreds to thousands of kilometres of habitat unsuitable for the organisms that are adapted to chemosynthetic conditions. An additional challenge is that most of the organisms that inhabit chemosynthetic environments are either sessile (being attached to a substratum) or show limited mobility in their adult life; they rely solely on planktonic propagules, mainly larvae (Mills et al. 2009) to maintain existing populations and to colonise newly opened areas (e.g., after an eruption at a vent or when a whale lands on the ocean floor). Given the patchy distribution and ephemeral nature of their habitat, adaptations during larval life can have pronounced implications for the success of these species. Yet, our knowledge of the larval ecology of these species, and of deep-sea species in general, remains extremely limited (reviewed in Young 2003; Mills et al. 2009).

Most of our current understanding of larval ecology is based on species that inhabit hydrothermal vents, and information from other chemosynthetic habitats is sorely lacking. Larval populations are being increasingly sampled to assess their abundance and distribution relative to the hydrothermal vent where they most likely originated. In general, larvae are found in greater abundance near the ocean floor than near the plume at hundreds of metres above the bottom, suggesting that they may be dispersing along the ocean floor, taking advantage of the along-axis currents there. Although these types of study were initiated in the 1990s (e.g., Kim et al. 1994; Kim & Mullineaux 1998), they accelerated in the 2000s. However, they have only focused at a handful of sites on the Juan de Fuca Ridge (Metaxas 2004), the East Pacific Rise (Mullineaux et al. 2005; Adams & Mullineaux 2008), and the mid-Atlantic Ridge (Khripounoff et al. 2001, 2008).

Similar studies were initiated in the 2000s and are ongoing at vents in Lau Basin and volcanically active seamounts on arcs in the southern (Kermadec) and western (Mariana) Pacific. These studies suggested that hydrodynamics can provide a mechanism of both larval retention to re-seed existing populations, as well as along-axis transport and dispersal to colonise newly opened areas within hundreds of kilometres. Larval abundance in cold seeps has been measured for two species in the Gulf of Mexico (Van Gaest 2006; Arellano 2008), indicating that, unlike most species at vents (except some crustaceans), larval migration from the seep of origin to surface waters likely occurs.

Larval colonisation is better understood and has received more attention than larval dispersal. Since 2002, studies have focused on vents, seeps, whale and wood falls. In all habitats, larvae of different species settle and colonise areas in a particular sequence that appears to be related to chemical and biological cues of the environment.

We now know that colonisation and succession at vents can be quite rapid and communities can recover from catastrophic disturbances within 2–5 years (Shank et al. 1998). It appears that the spatial and temporal patterns of colonists are primarily related to the physicochemical environment and secondarily to biological interactions. However, the evidence on the latter is still scant, and both experimental manipulations and numerical modelling are being used increasingly to address this gap (Neubert et al. 2006; Shea et al. 2008; N. Kelly, personal communication).

The largest gap in our understanding of larval life is the factors that affect larval growth and development in all chemosynthetic environments. The main challenge is larval rearing for species that inhabit deep-water environments with very particular chemical and physical characteristics. Only a few studies have succeeded in rearing larvae of only a handful of species from vents (Marsh et al. 2001; Pradillon et al. 2001), seeps (Young et al. 1996; Van Gaest 2006; Arellano & Young 2009), and whale falls (Rouse et al. 2009), but none were successful in following larvae through to the end of that life stage. Larval rearing in situ has been attempted and has been partly successful (Marsh et al. 2001; Pradillon et al. 2001; Brooke & Young 2009).

A key unknown aspect of the question of larval development is the duration of the larval stage, the period that larvae spend in the water column, and thus their potential dispersal distance. Some specific exceptions include experiments to estimate dispersal time in the vent tubeworm *Riftia pachyptila* (Marsh et al. 2001), the whale-fall polychaete *Osedax* (Rouse et al. 2009), the vent gastropod *Bathynerita* (Van Gaest 2006), and the mussel *Bathymodiolus* (Arellano & Young 2009). Other major gaps in knowledge include larval growth rates in relation to temperature, pressure, and food availability; the cues that induce them to stop swimming and settle onto a suitable habitat (or avoid an unsuitable one); and the role of larval behaviour in vertical positioning while in the water column or near bottom. Increased knowledge of larval connectivity and colonisation processes will add conceptual understanding of reducing ecosystems as metapopulations (Leibold et al. 2004; Neubert et al. 2006) and their response or resilience in the face of natural and anthropogenic disturbance (Levin et al. 2009b).

Understanding Remote and Dynamic Ecosystems

Discovery of deep-water chemosynthetically driven communities is relatively recent. Thus the investigation of these habitats has a very strong exploration component. Biologically, the unknowns exceed the knowns. How many species are there? What is their distribution and why? How do species reproduce, disperse, and colonise new sites? What are the evolutionary history and phylogenetic relationships of species from different chemosynthetic habitats? The remoteness and abrupt topography of deep-water chemosynthetic ecosystems make observation, sampling, and experimentation difficult in these habitats. There is a strong need for international collaboration and sharing of resources, which has been accomplished through projects such as the Census, contributing greatly to our knowledge of chemosynthetically driven ecosystems. The ongoing learning process allows identifying gaps and provides a driver for science to transform the unknowns into knowns (Gomory 1995; Marchetti 1998).

Chemosynthetic research depends on our capacity to sample, observe, and experiment at great depths in "extreme" physicochemical conditions. Therefore research in chemosynthetic ecosystems closely follows technological developments. For example, the use of submersibles and ROVs made possible direct observation and in situ precise sampling of the ecosystem. AUVs are useful for investigation of regions that are not accessible from the surface (i.e. oceans under ice) and in revolutionising the efficiency of deep-sea hydrothermal exploration (German et al. 2008a; Sohn et al. 2008). Deep-towed sidescan sonar instruments can produce detailed acoustic images of the deep seafloor.

Submersible-mounted multibeam bathymetry can achieve centimetre-scale resolution maps of the sea floor. High-resolution, high-definition cameras are being used to produce photo-mosaics of remote habitats, providing a comprehensive overview of the habitat and allowing for first interpretations of the relationships between habitat and fauna. New types of biogeochemical sensor module and incubation instrument allow quantification of the transport of energy and the benthic community activities in situ (Boetius & Wenzhöfer 2009). Technology is also advancing rapidly in laboratory techniques, for example molecular biology, which has aided our understanding of processes that, until recently, were hidden from our senses. However, are there limits to knowledge? This question can be considered in terms of different timescales. At the ecological timescale, one of the major limits to knowledge is imposed by society itself. The first barrier is one of economics and human resources.

In the case of the discrete and dynamic deep-water chemosynthetic ecosystems, some aspects might be unknown and unknowable at any given time. German & Lin (2004) have estimated that, for fast- and intermediate-spreading ridges, there should be a volcanic eruption approximately every 50 years for any given 100 km of ridge section. Slow-spreading ridges may exhibit much greater irregularity (German

& Lin 2004). It would take the equivalent of one 1 to 2 year expedition to explore all of the southern MAR from north to south. Taking into account the episodic volcanic activity of mid-ocean ridges, we would expect three or four major new eruptions to have occurred over the time it would take to explore the whole 7,000 km of the MAR. By that reckoning (approximately one new vent area on the MAR each year), our current rate of discovery (nine sites on the northern MAR since 1986) is not even keeping pace with the rate of new production. Here, the limit to knowledge is a consequence of the dynamic characteristics of mid-ocean ridges.

Cold seeps at passive margins are more stable ecosystems than vents (Sibuet & Olu 1998), but those on active margins (e.g., Chile, Costa Rica) are subject periodically to some of the most violently destructive forces on Earth. We can study these systems in their current "dormant" state and investigate their geophysics and geochemistry interactions and associated fauna. However, the episodicity of the great earthquakes (an approximate 100 year cycle predicted for the magnitude 9.5 great earthquake) may render the responses to such events impossible to observe and the extent to which the associated organisms are impacted by or even anticipate major tectonic events unknowable. Similar issues apply to stochastic, discrete, and ephemeral "habitats" such as whale falls and large organic falls. The life cycle of vents on fast-spreading ridges (i.e. East Pacific Rise) or cold seeps on active margins is equivalent, in time, to the construction of a cathedral in the Middle Ages (approximately 100 years), whereas on slower-spreading ridges they can extend to more than 10,000 years (Cave et al. 2002), longer than the oldest known European prehistoric constructs such as Stonehenge in the southern United Kingdom. These are unknowables for ChEss.

Furthermore, the fauna from deep-water vents, seeps, and whale falls is often new to science. Although the diversity of megafauna is low and the new species are described at a pace that keeps with discovery, this is not the case for the more diverse meiofauna, where up to 90% of the species collected can be new to science. The inability to keep up the rate of taxonomic identification is increasingly affected by the continuous decrease in taxonomic expertise among the new generations of marine biologists.

Marine Biotechnology and Microbiology

We are in charge of graduate school education on microbiology, molecular biology and epizootiology for aquatic organisms. One of our goal is to advance the sciences regarding comprehensive physiology and ecology for aquatic organisms by molecular and cell biological techniques. Another goal is to prevent and control infectious diseases of cultured organisms.

Marine Microbiology

In this field, main subjects are taxonomy, ecology and host-microbes interaction of marine microbes by using physiological, biochemical and molecular biochemical techniques. We are also searching for useful gene resources of marine bacteria as subject of applied Microbiology.

Main researching themes: Identification and classification of microbe of disease associate with marine animals and plants. Searching for symbiotic marine bacteria associated with marine animals. Molecular ecology of pathogenic bacteria and symbiotic bacteria associated with marine animals. Searching for sensing factors possessing pathogenic bacteria and symbiotic bacteria associated with marine animals to host animals.

Searching for valuable enzyme-coding genes from marine environmental DNAs.

Marine Molecular Biology

Laboratory of Maine Molecular Biology deals with genetic engineering for various functional proteins and enzymes from marine animals and plants, and their applications to industrial uses.

Current Research Subjects are as Follows

Isolation and cDNA cloning of industrially useful enzymes, cellulose, alginate lyase, mannanase, protease, from marine invertebrates such as scallop, abalone, and sea urchin.

Mass production of recombinant enzymes of marine organisms by using bacterial and insect cells. Production of protoplasts from brown seaweeds by using polysaccharide-degrading enzymes from marine invertebrates.

Isolation and characterisation of enzymes related to growth and differentiation of brown seaweeds. Study on regulatory mechanism of troponin from scallop adductor muscles.

Aquatic Epizootiology

Outbreaks of diseases are one of the most serious problems on aquacultured organisms for securing of food source. In our laboratory, we focus on infectious diseases of fish and shellfish, especially diagnosis and treatment of the diseases, moreover molecular and microbiological studies on the causative agents of the diseases. Also, we perform studies on prevention of diseases for aquatic organisms.

Marine Bioresource Chemistry

The laboratory consisted of two fields, Marie Biofunctional Chemistry and Marine Bio-analytical Chemistry, is engaged in the education for the graduate school.

The lectures on the understanding of unique life in marine organism, characterisation of unique compounds for marine life, functionalities of newly found compounds from marine organism, and maximal utilisation of marine resources as food, nutraceutical, and pharmaceutical compounds.

Biofunctional Chemistry

Marine resources contain a variety of functional compounds. We study on the chemical properties, nutritional effects, biological functions, health benefits, and utilisation of these marine bioresources, and their application to food materials, nutraceuticals and medicines. Absorption, metabolism, nutrigenomics, and molecular biology of marine functional compounds are also our main research themes.

Our research project also includes stabilisation of functional components in food systems and their enzymatic modification for higher biological activities.

Bio-analytical Chemistry

This field has focused on teaching and researching on analytical chemistry stereochemistry and biochemistry of marine organic compounds, particular lipids and lipid related compounds.

Current research activities: stereochemistry of glycerolipids in fish, algae and bacteria; stereospecificity of lipid related enzymes such as lipases; precision analysis of fish oils; analysis, distribution and function of marine organic compounds with unusual structures.

Biomolacular Chemistry

The field is engaged in the education and research on the understanding of unique properties of various compounds in marine resources, and the development of the purification, analysis separation, isolation, and evaluation systems.

Marine Products and Food Science

This laboratory provides research training for graduate students and also carries out various other activities which include development of new theories and highly improved technologies for effective utilisation of marine bioresources as foods and other value added or highly consumed products.

Graduate students are trained in specialised fields and research along their priorities of interest in interdisciplinary fields including chemistry, biology, biochemistry, microbiology, and also production engineering in the related fields.

Marine Bioresources Utilisation

To explore the useful potentials as health beneficial, drugs and for other purposes materials from marine bioresources are the mission of this field. We are identifying some beneficial enzyme inhibitors such as against glucosidase, urease, glucuronidase and so forth from algae. Our current topic is the discovery of a beneficial therapeutic potential against tumour of marine phospholipid liposomes which act as an immune system booster in combination with glucan.

Food Biochemistry

With a view to utilise the fish and shellfish at a high-degree, we educate and study on the biochemical properties of muscle proteins and endogenous enzymes, such as proteinases and transglutaminases, which are closely related to the freshness and quality of seafood. We also work on the structure-activity relationship of fish and shellfish proteins by using molecular biological and genetic techniques. Based on these studies, we attempt to develop a novel preservation methods and to make new seafood products by control of the endogenous enzymes in fish and shellfish.

Food Functional Chemistry

This field covers the applied science and technology for high utilisation of marine bioresource proteins as food materials and bioproducts. In recent topics are as follows: (1) Applied protein chemistry: Attempts to improve protein-functionality (food process characteristics and biofunctions) using molecular modification. (2) Seafood allergy: Identifying allergenic proteins in seafood and effect of food processing on the allergenicity. (3) Applied marine enzymology: Search for new enzymes and enzyme-inhibitors from unutilised marine bioresources and developing their application.

Food Safety

This field includes food preservation and food hygiene, and deals with following subjects to assure wholesomeness of marine foods: (1) specification of causative microorganisms and chemical substances of food deterioration and food-borne diseases, (2) analyses of mode of the actions and development of controlling methods with , cellular biological and molecular biological techniques, and (3) classification and beneficial utilisation of food microorganisms.

Chair of Marine Biosafety Science and Technology

The total safety of marine organisms from the production process to consumption is an urgent social need. This laboratory is designed to educate students and conduct research on both Marine Environmental Biosafety Management and the Marine Life Safety and Quality Management of Food.

Marine Environmental Biosafety Management

The main objectives of this field are the management of aquaculture environment and production of safe aquaculture feed from fisheries wastes for sustainable aquaculture. We provide education and conduct research on the separation science and technology for the removal of toxic substances from aquatic environments and feed ingredients.

Food Chain Safety and Quality Management

To ensure the safety and reliability on food chain, we have to continue the analysis and control of biological, chemical and physical hazard factors during growth, harvesting, processing, distribution and consumption of food. We try to systematize the safety management of food chain. The other target of this chair is to train persons who have highly expert knowledge about international assurance system of food.

Microbiology

Marine microbiology is the study of marine life forms that cannot be seen with the naked eye, including viruses, bacteria, fungi, protozoa and microscopic algae. Representatives of each group have been found in almost every marine habitat examined, from tropical coral reefs to the Antarctic and the greatest depths of the ocean. Microbes occur in the water column and sediments, on the surfaces of marine plants, animals and inanimate objects, in the intestinal tracts of fishes and invertebrates and even within the tissues of other living things. There is a staggering diversity of forms and modes of life among marine microbes. Some require light for growth, many require the saline nature of sea water, while others can survive without oxygen.

Because of their small size, the importance of microbes to life in the oceans is often not appreciated. Knowledge of their activities assumes a new importance when one considers that each millilitre of sea water may be the home for millions of individuals, and each gram of sediment may support thousands of millions of microbes. Microbes play a crucial role in ocean food webs, which support the diversity of life in the sea. Scientists therefore study the microscopic plants that fix carbon dioxide, the bacteria which grow on the myriad of organic, inorganic, dissolved and particulate chemicals in the sea, and the protozoa which feed on them. In this way knowledge is gained about how each life form contributes to all levels of the food web.

Other microbiologists investigate the unique ways in which microbes change chemicals from one form to another, and the importance of this to the health of the oceans. For example, particular kinds of microbes carry out some steps in the cycling of nitrogen from a dissolved gas in the sea to proteins in plants, animals and microbes, to inorganic forms and finally back to a gas. Maintenance of this cycle is essential for life in the oceans.

Many aspects of marine microbiological research have the potential to provide practical information and to form the basis of new biotechnological developments. Microbiologists studying diseases of marine organisms are providing disease diagnosis and control strategies to those attempting to grow marine animals such as prawns, molluscs and fish. New chemicals with potential industrial applications, such as friction-reducing lubricants on the surfaces of some fish, have been found to be produced by marine microbes. Major efforts are also being put into the search for new pharmaceuticals from marine microbes, such as antibiotics and anticancer agents.

The incredible diversity of microbes, their wide distribution in marine habitats, metabolic capabilities and intimate involvement in processes occurring in the oceans ensures lifetimes of discovery for scientists.

Research into the activities of microbes involves opportunities for collaboration with scientists from disparate disciplines such as botany, geology, zoology, chemistry and marine biology. Field work and laboratory analysis using traditional microbiological methods as well as sophisticated techniques such as DNA technology are all tools of trade for the marine microbiologist. Marine microbiologists require tertiary qualifications and are employed by universities, a variety of state and federal departments and instrumentalities, and by private industry.

Improving Human Life, Beginning with the Basics

If we can master the fundamental biological processes that shape life, we will possess the basic information needed to address diseases that afflict individuals at every stage of their lives. The University's initiative, "The Basics of Life," expressly capitalises on this concept. An infusion of private funds will catalyse the investment of others, including government and industry, and enable us to build on our leadership in key areas of basic and clinical research, accelerating the development of lifesaving therapies. In the process, we will establish a powerful model for research universities across the nation.

Creating Knowledge To Save Lives

The premise behind the University of Virginia's initiative, The Basics of Life, is compelling. If we can understand fundamental issues—such as how cells become specialised and form tissue, how they communicate with each other and respond to their environment, and how they migrate through the body—we will gain the insight needed to cure a panoply of diseases that affect human beings.

When lives are in the balance, basic information is crucial. Only through detailed knowledge of the formation, maintenance, degeneration, and regeneration of molecular networks, tissues, and organs can we prevent birth defects; control the rampant tissue growth associated with cancer; slow the deterioration that characterises aging; and facilitate the repair, regeneration, and replacement of injured tissues.

Fundamental advances in biological research promote the development of diagnostic tools and therapies throughout the entire life span. Breakthroughs at any one stage of human development can affect treatment at every other stage. At U.Va., we propose to reorganise our research efforts in new and creative ways so that we can more effectively apply our Basics-of-Life strategy to cure or prevent disease, while developing collaborations and partnerships worldwide to extend this model to other major research universities. We will create University-wide initiatives through three institutes and a research centre: the Morphogenesis and Regenerative Institute, the Virginia Institute for Clinical and Translational Research, the Institute on Aging, and the Pediatrics and Birth Defects Research Centre.

Building from a Position of Strength

Thanks to strategic investments by the University and the Commonwealth during the last five years and the extraordinary success of our faculty in securing outside research funding, we have assembled many of the elements needed to assume national leadership in driving lifesaving breakthroughs in diagnosis and treatment of cancer, birth defects, diabetes, and heart disease through advanced research in biology and medicine.

Morphogenesis and Regenerative Medicine Institute

Founded in 2002, the multidisciplinary Morphogenesis and Regenerative Medicine Institute was formed explicitly to pursue one of the most promising lines of inquiry in biology and medicine: expanding our knowledge of how cells differentiate and organise themselves so that we can harness these insights to regenerate defective or damaged organs. The potential of discoveries in this area is stunning, offering us the ability to repair spinal cord injuries, for instance, or restore lung tissue damaged by years of smoking or coal mining.

The institute brings together U.Va.'s world-leading programs in such fundamental biological processes as the migration of cells through the body, the mechanisms that enable cells to adhere to each other, and the signalling pathways that control the organisation of tissues. It also draws on our outstanding expertise in kidney and blood vessel development, among other areas. Virginia Institute for Clinical and Translational Research.

The newly created Virginia Institute for Clinical and Translational Research will play a critical role in extending the benefits of basics-of-life science to the diseases of adults. This research "translates" basic scientific discoveries into what Mr. Jefferson called "useful science" that has direct applications in patient care. Initially, we intend to combine our expertise in tissue development and diabetes to transform precursor cells in the pancreas to cells that can express insulin. The result of this dramatic breakthrough: a cure for Type 1 diabetes. In the process, we will extend our expertise to other areas, such as vascular disease and cancer, which can benefit from similar approaches.

Institute on Aging

The University's Institute on Aging brings together leading authorities in psychology, biology, medicine, and engineering from the United States and Europe to focus on mobility and independence, cognitive aging, age-related neurological disease, and other important issues. U.Va. scientists are responsible for one of the world's foremost longitudinal studies of aging, allowing medical researchers to compare declines in cognitive function with changes at the cellular and molecular level. These correlations provide the fundamental insights needed to pursue more effective strategies to bolster cognitive function and to prevent and treat such a Ge-related illnesses as Parkinson's and Alzheimer's.

Taking the Next Step

The University has put in place many of the elements needed to realise the promise of its basics of life strategy, and the University's Board of Visitors has committed to strengthen scientific and medical research even further. In recent months, the University has recruited several distinguished senior researchers in neuroscience, cell biology, and epidemiology, all areas that are vital to our knowledge of cell, tissue, and organ development. To complement the exciting work of these institutes, we will establish a Pediatrics and Birthdefects Research Centre that combines

University expertise in the Schools of Medicine, Education, and Law. The Pediatrics and Birth Defects Research Centre will apply insights from basic research to address major threats to the health of the fetus, optimise the care of the newborn, and address the treatment of children with acute brain injury or spinal cord injury. The work of the centre will gain additional momentum from the construction of a new U.Va. Children's Hospital.

There is, nonetheless, much more work to be done. An infusion of significant private funding will jump-start our efforts ($300 million in endowment will be required to support properly the four world-class centres). This support will enable us to make strategic additions to our faculty, develop advanced research instruments and facilities, and assemble multidisciplinary research groups that can speed the translation of basic research into real-world clinical applications. With these new resources, the University of Virginia can become the place where birth defects are prevented, where we learn how to prevent cancer cells from moving to lymphatic nodes, where both Type 1 and Type 2 diabetes are cured, where early heart attacks are prevented, and where cognitive and physical functionality are sustained in old age.

In short, by enabling us to complete and energize this unique network devoted to biology and medicine over the course of the human life, these additional funds will accelerate the flow of lifesaving therapies to those who need them most. Our goal? To improve human life, beginning with the basics.

Molecular Biology

Genome sequence comparison has been an important method for understanding gene function and genome evolution since the early days of gene sequencing. Alignment of DNA sequences is the core process in comparative genomics. In recent years, an important new sequence-analysis task has emerged: comparing an entire genome with another. Several powerful alignment algorithms have been developed to align two or more sequences.

MUMmer

MUMmer is a system for rapidly aligning entire genomes, whether in complete or draft form. MUMmer can also align incomplete genomes; it can handle thousands of contigs from a shotgun sequencing project, and will align them to another set of contigs or a genome using the NUCmer program included within the system. If the species are too divergent for a DNA sequence alignment to detect similarity, then the PROmer program within the environment can generate alignments based upon the six-frame translations of both input sequences. The original MUMmer system, version 1.0, was described in a 1999 Nucleic Acids Research paper. Version 2.1 appeared a few years later and was described in a 2002 Nucleic Acids Research paper, and the most recent version MUMmer 3.0 was described in a 2004 Genome Biology paper.

BLAT

BLAT (The BLAST-Like Alignment Tool) is a new tool for sequence alignment, which is similar in many ways to BLAST. The program rapidly scans for relatively short matches (hits), and extends these into high-scoring pairs (HSPs). However, BLAT differs from BLAST in several significant ways. Specifically, where BLAST builds an index of the query sequence and then scans linearly through the database, BLAT builds an index of the database and then scans linearly through the query sequence. Where BLAST triggers an extension when one or two hits occur in proximity to each other, BLAT can trigger extensions on any number of perfect or near-perfect hits. Where BLAST returns each area of homology between two sequences as separate alignments, BLAT stitches them together into a larger alignment. Both the client/server and the stand-alone can do comparisons at the nucleotide, protein, or translated nucleotide level.

MEGABlast

Mega BLAST uses the greedy algorithm of Zhang et al. for nucleotide sequence alignment search and concatenates many queries to save time spent scanning the database. This program is optimised for aligning sequences that differ slightly as a result of sequencing or other similar "errors". It is up to 10 times faster than more common sequence similarity search and alignment programs and therefore can be used to swiftly compare Microbial population genetics is a rapidly advancing field of investigation with relevance to many areas of science. The subject encompasses theoretical issues such as the origins and evolution of species, sex and recombination. Population genetics lays the foundations for tracking the origin and evolution of antibiotic resistance and deadly infectious pathogens and is also an essential tool in the utilisation of beneficial microbes.

Written by leading researchers in the field, this invaluable book details the major current advances in microbial population genetics and genomics. Distinguished international scientists introduce fundamental concepts, describe genetic tools and comprehensively review recent data from SNP surveys, whole-genome DNA sequences and microarray hybridisations. Chapters cover broad groups of microorganisms including viruses, bacteria, archaea, fungi, protozoa and algae. A major focus of the book is the application of molecular tools in the study of genetic variation. Topics covered include microbial systematics, comparative microbial genomics, horizontal gene transfer, pathogenic bacteria, nitrogen-fixing bacteria, cyanobacteria, microalgae, fungi, malaria parasites, viral pathogens and metagenomics.

An essential volume for everyone interested in population genetics and highly recommended reading for all microbiologists.

The higher taxonomic groups within prokaryotes are presently distinguished mainly on the basis of their branching in phylogenetic trees. In most cases, no molecular, biochemical or physiological characteristics are known that are uniquely shared by species from these groups. Analyses of genome sequences are leading to discovery of novel molecular characteristics that are specific for different groups of bacteria and archaea and provide more precise means for identifying and circumscribing these groups of microbes in clear molecular terms and for understanding their evolution. These new approaches and their limited applications for clarifying microbial systematics are described here. Because of their taxa specificities, further studies on these newly discovered molecular characteristics should lead to discovery of novel biochemical and physiological characteristics that are unique to different groups of microbes.

Microbes are ubiquitous in the world in which we live. With the development of high throughput DNA sequencing technology, there has been an explosion of DNA sequence data on microbes. The major aim of future microbial genomics will be to identify the functional significances of individual gene and genomic fragments and to use the information to help improve human health and promote our society development. One current major undertaking to understand genomic information is the comparative analyses between genomes that are not only distantly related, but also closely related ones.

Such comparative analyses between genomes that have diverged at different evolutionary time scales allow us to extract different types of information about biological functions and evolutionary processes. We review the tools and databases that have been established for comparative analyses of microbial genomes and discuss the implications of such analyses on our understandings of the common properties of life, the extent of genome plasticity and diversity within and between species, the processes and mechanisms underlie the observed genome diversity, and the origin and evolution of life.

Horizontal gene transfer, as a major force in shaping bacterial gene content, has gained incredible attention over the last decade. Along with the fast growing bacterial genome sequence data, there have been an increasingly large number of studies focused on horizontal gene transfer. The studies have been gradually transformed from identifying individual genes that have been horizontally transferred to assessing the general patterns of horizontal gene transfer and evaluating the systematic consequences of massive gene transfers.

The rates of gene transfers have been measured by various methods such as parsimony and maximum likelihood methods. Different phylogenetic methods were applied to a variety of data sets to assess whether there exists a congruent and meaningful bacterial tree. Even though some consensus has been reached, many contradictions have emerged and need to be solved in future studies. Our goal here is to review recent studies on this subject with an emphasis on the emerging patterns in horizontal gene transfer research.

Population genetics examine variation in genes among a group of strains of a particular species. Its major theme is to look at how different environmental factors and selective pressures can affect the distribution of genes and alleles. In this chapter *Yersinia pestis* was employed as an example to illustrate how the techniques are used for population genetic studies and how the achievements of these kinds of studies can be used for rapid identification and tracing the origin of pathogenic bacteria. New emerging techniques, including high throughput sequencing technologies, will give us unprecedented opportunities to understand microevolution and pathogenesis of bacterial pathogens.

Symbiotic nitrogen-fixing Rhizobia are of global significance, both in terms of their ecological relationships and their importance as an environmentally benign source of nitrogen for crop plants. These bacteria are capable of forming mutualistic relationships with a variety of legume hosts, where they convert atmospheric nitrogen (N_2) to ammonia that is used to help meet the nitrogen needs of the host plant. In this chapter, we review our current understanding of the population genetics of this diverse group of bacteria. First we briefly describe the various types of genomic architectures that are found within rhizobial species. Next we outline the phylogenetic relationships among the recognised taxa. We then summarise the results of studies that have examined patterns of molecular genetic variation in rhizobial populations at local, regional, and global levels.

The results of these studies have revealed several important insights, including the existence of extensive diversity within, as well as significant genetic differentiation between local and regional populations. The results also provide evidence for long-distance gene flow between continental populations. Several studies have also indicated the existence of low-to-intermediate levels of recombination within rhizobial populations. Fine-scale studies of specific genomic components (e.g., symbiotic plasmids) have also shown that certain genomic elements appear to be more prone to recombination than others.

The results also showed that the different loci responsible for the development of the symbiosis appear to be under different forms of selection. Because most the population studies to date have focused on strains from root nodules, surprisingly little is known on the population genetics of the more numerous non-nodulating soil rhizobia. Future efforts to characterise these populations should significantly enhance our ability to manipulate rhizobial populations in agricultural ecosystems. Cyanobacteria are a group of ecologically diverse photosynthetic bacteria. Because niche differentiation is ultimately the product of differences among individuals within populations, understanding the evolutionary origins of this diversity ultimately requires a population genetics perspective. This chapter considers the current state of our understanding of the mechanisms that generate variation in cyanobacteria, the distribution of this diversity and its potential functional importance, with an emphasis on recent work from our laboratory regarding diversity within and among populations of the cosmopolitan, multicellular cyanobacterium, *Mastigocladus laminosus*. It concludes with an appraisal of the potential to take a population genomics approach to address fundamental questions regarding the nature of adaptive variation and niche differentiation in these microorganisms.

Algae are a highly diverse group of protists, ranging from simple, unicellular organisms to complex, multicellular entities with a range of differentiated tissues and distinct organs. They are found among diverse aquatic ecosystems and play important roles by supplying carbon and energy as well as providing habitat to other members of the biological communities. Some algae cause significant environmental and health problems. However, despite their importance, relatively little is known about this group of organisms.

The first section of this chapter briefly summarises our current understanding of the diversity and evolution of the algae. The second section reviews the molecular markers used to address algal population genetic issues. The third section summarises the population genetic analyses of three algal groups: the dinoflagellates, the diatoms and the haptophytes. With the application of genomics information and technology, the field of algal population genetics is approaching an exciting period of development and expansion.

Mutualisms are reciprocal exploitations that nonetheless increase the fitness of each interacting partner. Two groups of fungal mutualists of plants, epichloë endophytes of grasses and arbuscular mycorrhizal fungi, were selected as focal systems to discuss population-level processes that contribute to the establishment and maintenance of mutualistic interactions. These two classes of fungal cooperators of plants are subject to different and often conflicting selective pressures and represent distinct trajectories of mutualism evolution. Yet, in both cases population structure of symbionts is a source of information critical for understanding how these fungi interact with their hosts.

This article reviews the more common DNA-based markers that are used to genotype fungi, as well as other eukaryotes, and presents several examples of their application to elucidate the population genetics of mammalian pathogenic fungi. The most common fungal infections are briefly summarised to illustrate the issues that are routinely addressed by population genetics approaches. This exciting area of research is continually growing as the methods become more accessible and medical mycologists recognise the value of studying natural isolates of pathogenic fungi.

Genetic diversity, population structure, and evolutionary history of malaria parasites are among the key factors that will influence our ability to identify genes contributing to drug resistance, parasite development, and disease pathogenesis. These factors also have an impact on vaccine and drug development, parasite source tracking, as well as the formulation of other disease prevention and control measures. For example, a highly polymorphic parasite population will contain ample genetic diversity capable of generating drug resistance genotypes at an accelerated rate; while the presence of homogeneous parasite populations should aid in the development of an effective malaria vaccine. Malaria research in the post-genomic era offers many new tools for use in population genetics analyses.

Many viral pathogens, especially those with an RNA genome, are characterised by their high mutation rates and large population sizes. These features are responsible for the high levels of genetic variation usually found in viral populations and for their rapid response to different selective challenges encountered during their infection and transmission processes. They are quantitatively and qualitatively so different from most other organisms that special models and concepts, such as the quasispecies model, have been developed to better describe the evolutionary dynamics of viral populations. Here, we review these and other salient features of viral pathogens, with a species emphasis on RNA viruses. We describe how population genetics theory provides an adequate framework for analysing and interpreting genetic variation in viral populations, for understanding their dynamics, and to study adaptive processes occurring therein. However, not all the evolutionary changes observed are due to the action of positive selection and this is not an all-mighty agent of evolutionary change. We finish this review by introducing a recently developed framework for integrating the evolutionary and epidemic behaviour of infectious organisms, known as phylodynamics, which is especially well-suited for fast evolving organisms such as viruses. Microbial population genetics examines the spatial and temporal patterns of genetic variation across diverse geographic scales and ecological niches. With the arrival of molecular biological techniques, the past 40 years have seen tremendous progresses in microbial population genetics. However, in recent years, the analyses of genetic materials directly from natural environments have revolutionised our approaches and understandings of the diversity, function, and inter-relationships among microorganisms in diverse natural ecological niches. The emergence and development of this expanding new field, that of metragenomics, has been primarily driven by technical and analytical methods developed from high throughput platforms for cloning, microfluidics, DNA sequencing, robotics, high-density microarrays, 2D-gel electrophoresis, and mass spectrometry as well as associated bioinformatics softwares.

In this chapter, I present a general overview of this exciting development, including the general approaches and common tools used for identifying the diversity and function of microorganisms in natural biological communities. Of special notes are the potential impacts of recent developments in single cell isolation, whole-genome amplification, pyrosequencing, and database warehousing on our understanding of microbial population structures in nature. These exciting developments are bringing significant opportunities as well as new challenges to the field of microbial population genetics.

Epigenetics

Epigenetics is the study of changes in gene expression caused by mechanisms other than changes in the DNA sequence. Epigenetics is a rapidly advancing field with an increasing impact on biological and medical research.

The editors of this book have assembled top-quality scientists from diverse fields of epigenetics to produce a major new volume. Comprehensive and cutting-edge, the 26 chapters in this book constitute a key reference manual for everyone involved in epigenetics, DNA methylation, cancer epigenetics and related fields.

Topics include: early life environment, DNA methylation and behaviour, histone acetyltransferase biology, transgenerational epigenetic inheritance, mammalian X inactivation, epigenetic memory in plants, polycomb group regulation, centromeres and telomeres, DNA sequence contribution to nucleosome distribution, macrosatellite epigenetics, histones, cell-fate specification and reprogramming, DNA methylation in cancer, variant histone H2A and cancer development, RNA modification, paramutation in plants, DNMT3L dependent methylation during gametogenesis, non-coding RNA, bisulphite-enabled technologies, rapid analysis of DNA methylation, microarray mapping, DNA methylation profiling, ChIP-sequencing, genome-wide

DNA methylation analysis, and epigenetics in maize. In addition there are useful chapters on bioinformatics in epigenomics, resources and tools for epigeneticists, and educational resources for epigenetics.

This up-to-date reference manual is an essential book for those working in the field and for scientists in other disciplines it represents a major information resource on the fascinating and fast-moving field of epigenetics.

The DNA molecule contains within its chemical structure two layers of information. The DNA sequence that bears the ancestral genetic information and the pattern of distribution of covalently bound methyl groups to cytosines in DNA. While the genetic information is similar in all tissues in the individual, the pattern of distribution of methylation across the genome is cell-type specific. DNA methylation is an important regulator of gene function. Recent data that will be discussed here that supports the hypothesis that DNA methylation is a reversible biological signal.

This expands the potential role of DNA methylation beyond embryogenesis to other time-points in life and to post mitotic tissues such as the brain. DNA methylation is proposed to act as a genomic response to both physical and social signals from the environment at different time points in life and to serve as a genomic memory of these exposures at different time scales, stably altering gene expression programming and thus modulating the physical and behavioural phenotypes to respond to these environments. It is hypothesized that DNA methylation provides within the structure of the DNA a dynamic interface between the changing world around us and the relatively fixed and stable genome.

A histone $(H3\text{-}H4)_2$ tetramer flanked by two H2A-H2B heterodimers form the core protein structure, around which DNA is wrapped. DNA and the histone octamer together form the smallest chromatin particle, the nucleosome. How intimately the DNA associates with the core histones and how tightly the nucleosomes are packed with each other is determined by a key post-translational modification of the histone proteins, namely acetylation. Histone acetylation was first discovered in the early 1960s.

After a dearth of progress, due to technical limitations, our knowledge of histone acetylation has exploded in the last fifteen years. Enzymes that catalyse acetylation of histones, the histone acetyltransferases, have been discovered, proteins associated with these have been identified and their preferences for specific histone residues have been determined. Importantly, we are gaining a better understanding of the relevance of histone acetylation in health and disease through the discovery of genetic mutations underlying human diseases in loci encoding histone acetyltransferases (HATs) and through examination of mouse strains deficient in specific histone acetyltransferases. Here we discuss the principles of histone acetyltransferase biology.

Epigenetic states are faithfully inherited through mitotic cell division, but are generally cleared and reset on passage through the mammalian germline. But this clearing of epigenetic marks is not always complete, leading to transgenerational inheritance of epigenotype. Transgenerational epigenetic inheritance has been demonstrated in several organisms, including mammals, and has been most comprehensively studied in mouse strains carrying variants of the *agouti* (A^{vy}) and *axin* ($Axin^{Fu}$) alleles. The most prominent feature of transgenerational epigenetic inheritance is its non-Mendelian nature: not all offspring that inherit the genetic locus also inherit the parental epigenetic state. Transgenerational epigenetic inheritance is emerging as an important facet of mammalian biology. It may underlie the etiology of human diseases that display complex patterns of inheritance, including diabetes, mental illnesses and autoimmune diseases. As variable epigenetic states can be inherited on an invariant genotype, epigenetic variation may provide a substrate for Darwinian selection that is independent of genetic variation.

X chromosome inactivation is the method of dosage compensation that has evolved to equalise expression of X-linked genes between female (XX) and male (XY) mammals. In somatic cells only one X chromosome is active; the second X in female cells is silenced early during embryonic development, a process that involves the co-ordination of multiple levels of epigenetic regulation to ensure stable chromosome-wide silencing. In this chapter we shall focus on the molecular mechanisms involved in X chromosome inactivation, discuss how the epigenetic marks are believed to elicit stable transcriptional silencing, and why X inactivation represents an excellent model system for studying. Polycomb-group (PcG) complexes are essential regulators of plant development. These multiprotein complexes repress gene expression by establishing and maintaining trimethylation of lysine 27 at histone H3, a modification that is associated with repressive chromatin. Recent studies have indicated that plant PcG complexes regulate key genes involved in responses to low temperature. Vernalisation is a long-term response to low temperatures whereby plants coordinate their seasonal flowering to occur after winter. In contrast, acclimation of plants to low temperatures, a key step in the establishment of frost tolerance, involves rapid activation of cold-acclimation genes. In this chapter, we describe the dynamics of PcG-mediated gene regulation underlying these two important agronomic traits that are triggered by low temperatures.

In eukaryotes, each chromosome has one centromere and its ends are protected by telomeres. The centromere is a specialised chromosomal locus that directs kinetochore assembly and provides the site for microtubule attachment, allowing accurate chromosome segregation during cell division. Despite the critical role centromeres play, centromeric DNA sequences are highly variable and not conserved between species. Increasing evidence, including the discovery of functional neocentromeres, suggests that centromere identity and function is epigenetically defined through the formation of a specialised chromatin structure.

This chapter reviews recent studies addressing the structural and functional characterisation of centromere chromatin, its assembly and propagation during cell division. Telomeres are specialised nucleoprotein complexes that protect the chromosome ends from degradation. In recent years, it has become increasingly clear that heterochromatic marks at telomeres act as negative epigenetic regulators of telomere elongation, repress recombination events at the telomere and are critical for maintaining telomere structural integrity. Recent research reporting telomeres being transcribed by RNA polymerase II to give rise to TERRA RNA, open up an additional level of regulation at the telomere. This chapter will discuss the links between the epigenetic status of telomeres, telomere function and telomere-length regulation, and the implications on cellular reprogramming, aging and cancer.

DNA in eukaryotes is efficiently and compactly organised into chromatin, the fundamental subunit of which is the nucleosome: approximately 150 bp of DNA spooled 1.65 times around a histone octamer. The location and density of nucleosomes play a role in regulating nuclear processes including transcription, replication, recombination, and repair. Mechanisms acting *in trans*, like ATP-dependent remodelers and cellular memory complexes, as well as *in cis* features intrinsic to the DNA sequence itself regulate the location and density of nucleosomes. Here, we review the three cis acting DNA sequence features that affect the distribution of nucleosomes: (1) two frameworks defining the relationship between the histone octamer and the underlying DNA sequence (nucleosome occupancy and nucleosome position, then statistical positioning and a nucleosome positioning code), (2) the organisation of DNA into the nucleosome core particle, and (3) specific DNA sequence features and DNA templates that promote or inhibit the formation of nucleosomes. We close by describing three computational algorithms trained on DNA sequence that have been used to predict nucleosome position and density. In summary, we hope to draw attention to multiple aspects of DNA sequence that specify organisation of sequence into nucleosomes and influence the distribution of nucleosomes in eukaryotic genomes.

The recent completion of several mammalian genome sequences makes obvious that we share a near-identical collection of genes. What defines us as human must therefore be encoded within regions of the genome where we differ, providing an added level of complexity that probably influences the spatial and temporal expression of genes. Most DNA sequence variation occurs within the repetitive DNA, once called 'Junk DNA' that accounts for at least half of the human genome, and evidence is mounting for its important role in genome function. Although some repeat elements are conserved to some extent between mammals, their precise copy number and genomic location typically are not.

In addition, some repeats are not conserved, including the large tandem repeats. This chapter focuses on two large tandem arrays in the human genome that can adopt quite different chromatin configurations as a result of epigenetic changes; one as a direct consequence of X chromosome inactivation and the other in the context of disease susceptibility. Both cases highlight how alternate packaging of these unusual DNA sequences probably results in differing functions. In each instance, common denominators are the acquisition of the epigenetic organiser protein CTCF and a distinct change in transcripts originating from the array.

In eukaryotes, the genetic material in the form of DNA is wrapped around histone proteins to form nucleoprotein filaments called chromatin. Histones help package the DNA to fit it inside the nucleus of each cell, which in turn regulates access to the genetic information contained within the DNA. Hence, all DNA transactions are likely to be affected by histone metabolism. Eukaryotes carry multiple histones genes that can potentially generate enormous quantities of histone proteins. When present in excess, the positively charged histones can potentially "stick" non-specifically to the negatively charged DNA and adversely affect processes that require access to DNA. Not surprisingly, aberrant histone stoichiometry, chromatin assembly or chromatin structure lead to genomic instability, which is characterised by the increased rate of acquisition of alterations in the genome and is associated with deleterious human conditions such as cancer and aging. Hence, histone synthesis is coupled to ongoing DNA replication and is regulated transcriptionally and post-transcriptionally. To further avoid the deleterious consequences of excess histones, a post-translational regulatory mechanism was described recently whereby excess histones are targeted for degradation by the ubiquitin-proteasome system. In this chapter, we discuss the causes and consequences of excess histone accumulation, as well as the strategies that cells have evolved to deal with them.

Cell-fate specification and stem-cell renewal are fundamental processes in the development of multicellular organisms. In both animals and plants, a key role for transcription factors in these processes has been established, but an accumulating body of evidence indicates that epigenetic regulation also plays a critical role. Once regarded as stable marks, all epigenetic modifications, including DNA and histone methylation, are now known to be reversible, and a cohort of enzymes that add or erase epigenetic marks has been identified. Throughout the life of an organism, the epigenome is dynamically modified, leading in turn to transcriptional changes and ultimately cell-fate specification. Here, I review the recent literature that has shaped this view. Plants are unique among complex organisms in having the ability to generate a whole new organism from a single differentiated cell. How plant cells retain this amazing capability while maintaining their cell identity is a mystery. I introduce a model system for study of cell-fate specification, stem-cell renewal, and reprogramming.

DNA methylation is an important regulator of gene transcription and a large body of evidence has demonstrated that aberrant DNA methylation is associated with unscheduled gene silencing, and the genes with high levels of 5-methylcytosine in their promoter region are transcriptionally silent. DNA methylation is essential during embryonic development, and in somatic cells, patterns of DNA methylation are generally transmitted to daughter cells with a high fidelity. Aberrant DNA methylation patterns have been associated with a large number of human malignancies and found in two distinct forms: hypermethylation and hypomethylation compared to normal tissue. Hypermethylation is one of the major epigenetic modifications that repress transcription via promoter region of tumour suppressor genes. Hypermethylation typically occurs at CpG islands in the promoter region and is associated with gene inactivation. Global hypomethylation has also been implicated in the development and progression of cancer through different mechanisms. This chapter will focus on DNA methylation as the major epigenetic mechanism involved in normal biological processes and abnormal events leading to cancer development. It will also focus on the interaction between DNA methylation and other epigenetic mechanisms.

The histone variants of the H2A family are highly conserved in mammals, playing critical roles in regulating many nuclear processes by altering chromatin structure. One of the key H2A variants, H2A.X, marks DNA damage, facilitating the recruitment of DNA repair proteins to restore genomic integrity. Another variant, H2A.Z, plays an important role in both gene activation and repression. A high level of H2A.Z expression is ubiquitously detected in many cancers and is significantly associated with cellular proliferation and genomic instability. This review summarises the current understanding of these variants and their functions, as well as their links to cancer development. Furthermore, the significance of dysfunction of these variants is highlighted with respect to their potential as biomarkers and as new targets for anticancer therapy.

A wealth of nucleobase and ribose modifications have been identified in multiple types of RNA including tRNAs, rRNAs, mRNAs, and small regulatory RNAs. Among them, 5-methylcytosine (m^5C) has been detected in rRNAs, tRNAs, and early reports have indicated its presence in mRNAs. Well established as an epigenetic mark in DNA, the prevalence and function of m^5C in RNA is either incompletely explored (tRNA, rRNA) or virtually unknown (mRNA, other noncoding RNA). Two eukaryotic m^5C RNA methyltransferases have been identified; however, their substrate specificity and biological roles are incompletely understood. With recent advances in bisulfite sequencing of RNA, comprehensive analyses to determine the occurrence and functions of m^5C in the transcriptome now appear feasible. In this chapter, we summarise the current knowledge in this field, focussing primarily on eukaryotic transcriptomes.

Paramutation is a fascinating phenomenon in which epigenetic information can be transmitted through *trans*-interactions between one allele of a gene to another allele or between homologous DNA sequences that establishes a state of gene expression that is heritable for generations. Paramutation was discovered in maize and similar phenomena have been described in other plants, fungi and animals.

In this chapter, we describe several classic plant paramutation systems and discuss recent advances that implicate a role for RNA and a number of components of an RNA-based transcriptional silencing pathway on paramutation.

DNMT3L (DNA methyltransferase 3 like) is member of the DNA methyltransferase family of enzymes responsible for the methylation of CpG dinucleotides. Biochemical studies have revealed that while DNMT3L lacks DNA methyltransferase activity, it can bind to and stimulate the activity of *de novo* DNA methyltransferases DNMT3A and DNMT3B. DNMT3L has also been shown to interact directly with chromatin via its plant homeodomain (PHD) like zinc finger domain. Studies in *Dnmt3L*-deficient mice have revealed that DNMT3L is essential for establishing correct methylation patterns at Retro Transposable Elements (RTE), unique loci and parentally imprinted genes in germ cells, and mice without DNMT3L are rendered infertile. Female *Dnmt3L*$^{-/-}$ mice have apparently normal meiosis but in male *Dnmt3L-/-* germ cells there is asynapsis of chromosomes, and "meiotic catastrophe". Dnmt3L was among the first mammalian genes shown to have a paternal effect, where the genotype of the father (*Dnmt3L*$^{+/-}$) affects sex chromosome aneuploidy in adult and embryonic offspring. This chapter will discuss the role of DNMT3L in establishing DNA methylation patterns during gametogenesis, as well as the proven and potential consequences of DNMT3L deficiency to fertility and somatic and germline genetic disease in light of the increasing evidence that epigenetic reprogramming is a dose sensitive and partially stochastic process.

In the past decade, we have become acquainted with an entire new world of fine regulatory control within cells governed by non-coding RNAs. Although we have not completely explored this world, we know from a few well studied examples that not only gene dosage but protein function also, is fine tuned by non-coding RNAs. It appears that proper cell function is largely dependent on non-coding RNAs, many of which were invisible to us until the advent of high throughput sequencing and tiling array technology. The realisation, that transcripts with no open reading frames are biologically active molecules with multiple functions including regulation of the expression of coding RNA, shifts the power from proteins to non-coding RNAs as key modulators of cellular function. In this review, we discuss the various classes of non coding RNAs and the mechanisms employed by them to achieve this status, with a particular focus on their ability to induce epigenetic modifications.

A large variety of methods to measure DNA methylation have been developed and used extensively over the last twenty years. These have been based on selective restriction digestion of methylated DNA, the capture of methylated DNA by methyl binding proteins or antibodies, or bisulphite conversion of DNA. However, all restriction enzyme based methods are dependent on available restriction sites for methylation-specific restriction enzymes and therefore cannot be used to analyse every CpG site in the genome. Although immunoprecipitation methods are not sequence specific, they are unable to provide methylation data at a single-base resolution. Therefore, both of these approaches are limited in their applicability. On the other hand, bisulphite conversion based methods allow methylation studies directed to any CpG site in the genome.

Sodium bisulphite treatment of DNA converts all unmethylated cytosines into uracil while methylated cytosines remain unchanged, thus transferring an epigenetic difference into a measurable genetic difference. A variety of downstream methods such as Polymerase Chain Reaction (PCR), sequencing, Single Nucleotide Polymorphism (SNP) genotyping and mass spectrometry can be coupled with bisulphite conversion. This chapter provides an overview of some of these methods, concentrating on bisulphite sequencing, methylation-specific PCR, pyrosequencing, MassARRAY EpiTYPER and Infinium Human Methylation27 Bead Chip. Considerations for assay selection and detailed protocols for each method are presented in the end of this chapter.

Methylation-sensitive high resolution melting (MS-HRM) is an inexpensive and robust closed tube screening methodology that enables rapid analysis of locus specific DNA methylation for multiple samples. MS-HRM is based on the differential melting behaviour of PCR amplification products derived from methylated and unmethylated templates after bisulfite treatment. The melting profiles of an unknown sample are compared to the melting profiles of standards with known DNA methylation levels. MS-HRM has the advantage of allowing ready distinction between homogenous and heterogeneous DNA methylation. Estimation of DNA methylation to quite low levels in a semi-quantitative manner is possible for homogeneously methylated templates where all the CpG sites are methylated. However for heterogeneous methylation, the formation of multiple heteroduplexes makes quantitation difficult. Digital MS-HRM utilises limiting dilution to amplify single templates enabling DNA methylation analysis at a single allele resolution and enables the quantitative analysis of both homogenous and heterogeneous methylation. The PCR products either generated by MS-HRM or digital MS-HRM can subsequently undergo Sanger DNA sequencing or pyrosequencing to investigate DNA methylation at individual CpG positions.

The location and density of nucleosomes in the eukaryotic genome plays a role in regulating nuclear processes including transcription, replication, recombination, and repair. Microarray mapping of nucleosomally protected DNA has emerged as a powerful, cost-effective, high-throughput method to analyse the relationship between nucleosome position and genomic regulation. In this chapter we discuss experimental considerations such as sample preparation and microarray design. In addition, two procedures are detailed: (1) Formaldehyde Crosslink and Harvest Cells, Isolate Nuclei, and MNase cut chromatin, and (2) Isolation of mononucleosomally protected DNA and fluorescent labelling of DNA for microarray hybridisation. With the information specified in this chapter, most any laboratory equipped for molecular biology with institutional or commercial access to microarray facilities should be should be able to map nucleosome position and occupancy.

DNA methylation plays an essential role in normal human development, where abnormalities in proper establishment and maintenance of DNA methylation patterns result in human disease. Many experimental approaches have been developed for assaying DNA methylation patterns, including enzymatic-based approaches. In this chapter, we highlight some of these approaches and describe their relative advantages and disadvantages. We also describe advances in microarray and sequencing technologies that have improved resolution of enzymatic-based methods, providing expanded coverage of CpG dinucleotides throughout the genome. These approaches are important tools in characterising the role of DNA methylation in genome organisation and function.

With the advances in traditional genetics failing to provide causal genes for many complex diseases, the focus of research is shifting towards determining the importance of gene-environment interactions, more specifically epigenetic regulation. Paramount to answering this question is the knowledge of which transcription factors bind to what sequence, as well as a detailed understanding of how the transcriptional state of a genetic sequence is epigenetically distinguished. The chromatin immunoprecipitation (ChIP) strategy has proven to be a powerful tool to investigate these mechanisms, but has been limited for a long time to single locus analysis. The recent emergence of next-generation sequencing (NGS) technology however has revolutionised the field of epigenomics and for the first time enabled unbiased genome-wide analysis of ChIPed DNA. Here we provide a detailed discussion and protocol on how to best perform ChIP for NGS analysis.

Over the past decades it has become ever more apparent that understanding the genome did not stop with unravelling the genetic code. Regulatory mechanisms are needed to determine which parts of the genome are active or inactive, and to form a memory system that can be passed on over multiple cell divisions for proper functioning of the cell. These mechanisms underpinning heritable gene regulation are encapsulated by the term "epigenetics" and include histone modifications, miRNA and non-coding RNA expression, and methylation of cytosine residues in DNA. In this review, we compare the various methods that can be used to analyse DNA methylation patterns throughout the genome, and discuss their advantages and disadvantages. In addition, we present a detailed protocol for genome-wide DNA methylation analysis based on the capture of methylated DNA using a methyl-CpG binding domain-based (MBD) protein combined with second generation sequencing.

Since biological phenotype and differentiation is regulated partially through CpG DNA methylation, and this mark is relatively easy to measure, genome-wide profiling of methylation landscape is a popular tool in epigenetic research. The burden then falls on bioinformatics to provide normalisation, quality control, and interpretation, while adopting to the vast number of versatile methods to interrogate methylation profiles. While creating a relational report of the results from either of these methods is critical for data to cross over from lab to lab, and from research to diagnostic and translation purposes, the methods are each introducing their unique bias to the data.

Ideally, one would hope all labs would adopt a single method, but since each method offers a unique advantage, such as price, sensitivity, or confidence, one has to overcome the disparity hurdle via bioinformatic tools. Thus identifying the methodological biases of each technique, and the way to compensate for those in the process of generating a universal CpG methylation prediction on a genome region is the ultimate goal we are describing here. Follow up of CpG methylation with other read outs, such as impact on gene expression or coincidence with locations in the genome, where allelic variation is associated with the investigated phenotype, are also key to proper interpretation of the results.

Maize has served as an excellent model for the study of epigenetic gene regulation for the past several decades. The pioneering work of maize geneticists like Barbara McClintock, Alexander Brink, Marcus Rhoades, and others led to the observation of many fascinating phenomena that were later demonstrated to be epigenetically regulated events. Since these observations were made, a great deal of progress has been made in determining the underlying causes of the phenomena, and many of these examples of epigenetic gene regulation are currently being used to elucidate the mechanisms of epigenetic heritability in plants. Today, these phenomena and the mutants that impact them serve as resources for studying how DNA methylation, chromatin structure, and small RNAs act to influence paramutation, gene silencing, and parent-of-origin dependent dosage compensation. The key attributes and potential contributions of each resource are discussed in the context of understanding the mechanisms and significance of epigenetic gene regulation in large, complex genomes.

Biological experiments are rapidly moving into the information age with the explosion of data generated from rapidly evolving technologies. Microarrays can interrogate many thousands to millions of loci within any one sample, while massively parallel sequencing platforms can essentially measure the entire genome. Keeping up with this rapid pace of data accumulation has required the development of online tools for the processing, analysis and annotation of data generated from large numbers of microarrays or next generation sequencing (NGS) experiments. I will cover a selected range of tools available to the biological researcher, from online discussion forums and blogs, to curated databases that store the data and associated annotations. I will then cover viewer tools that enable visualisation the annotations and finally review tools for assay design for follow up validation of NGS and microarray analyses. I will pay particular attention to DNA methylation and provide the reader with an insight into what is available online for the epigenetics researcher aiming to make sense of epigenomic data.

Epigenetics can appear as an impenetrable subject; not just to those encountering it for the first time, but to those within the field too. However, epigenetics, like any subject can be made easier to understand using a combination of clear language, creative illustrations and even animations and film clips. This chapter aims to point readers of all experiences towards helpful and easy-to-read resources that educate about epigenetics. It is split into two main sections, the first aimed at a lay audience including teachers and high school students and the second, at graduate and postgraduate students and beyond. Each section contains summaries of published articles and web sites. The chapter ends with a short section on epigenetic societies and research networks and a summary table of resources.

It is intended to provide a sample of some of the best short to medium length reviews on general topics within the field of epigenetics and while we cover a wide variety of themes, we apologise for any areas not covered. We cite the URLs of freely-available articles wherever possible, but many articles will require library access. We also urge readers to contact authors or publishers if they wish to distribute any of the articles for teaching purposes.

Systems Microbiology

Systems biology is the study of the dynamic interactions of more than one component in a biological system in order to understand and predict the behaviour of the system as a whole. Systems biology is a rapidly expanding discipline fuelled by the 'omics era and new technological advances that have increased the precision of data obtainable. A focus on simple single cell organisms such as bacteria aids tractability and means that systems microbiology is a rapidly maturing science.

This volume contains cutting-edge reviews by world-leading experts on the systems biology of microorganisms. As well as covering theoretical approaches and mathematical modelling this book includes case studies on single microbial species of bacteria and archaea, and explores the systems analysis of microbial phenomena such as chemotaxis and phagocytosis.

Topics covered include mathematical models for systems biology, systems biology of *Escherichia coli* metabolism, bacterial chemotaxis, systems biology of infection, host-microbe interactions, phagocytosis, system-level study of metabolism in *Mycobacterium tuberculosis*, and the systems biology of *Sulfolobus*.

This book is a major resource for anyone interested in systems biology and a recommended text for all microbiology laboratories.

Modelling methodologies in the life- and biomedical sciences are hampered by the complexity of the processes and systems at work. Modelling studies into prokaryotic systems require the elucidation of the mechanistic model. In this chapter we introduction modelling methodologies and discuss the problem of model (and parameter) inference. We comment on state-of-the-art research questions and provide a general discussion on how models can and should be used in order to better understand the structure, function and dynamics of biological systems. The aim is not to provide an introduction to modelling per se, but to provide readers with an overview on the available methodologies. The modelling approach chosen depends on the biological question at hand as well as a range of social factors.

The functional repertoire of an organism's metabolic network is closely linked to its phenotype and potential for utility in metabolic engineering applications. In this chapter, we discuss a systems biology view of *Escherichia coli* metabolism by integrating current genome-scale computational modelling approaches with available molecular genetics tools, as well as the experimental framework for metabolite and metabolic flux determination.

Bacterial chemotaxis is a paradigm for biological sensing and information transmission. The chemotaxis signal-transduction pathway allows cells to sense chemicals in their surroundings in order to regulate flagellated rotary motors, thus allowing them to swim towards nutrients and away from toxins. Importantly, cells are able to sense with remarkably high sensitivity over a wide range of chemical background concentrations. To make this possible, chemoreceptors do not signal independently but form clusters for amplification and integration of signals, as well as for adaptation to persistent stimulation.

While chemotaxis in *Escherichia coli* has been exceptionally well characterised, new experimental facts still require revisions of existing models and thus further increase our understanding of sensing and signalling in bacteria. Additionally, experiments on other bacterial species such as *Bacillus subtilis* and *Rhodobacter sphaeroides* indicate that bacteria other than *E. coli* can have substantially different and more complex chemotaxis pathways, which provides renewed challenges for experimentalists and modellers alike. Here we discuss our current understanding as well as the frontiers of bacterial chemotaxis research.

Microbial infections still cause around one quarter of all deaths globally, despite the advances that have been made in the treatment of infectious disease. The increasing occurrence of drug resistant pathogens, both old and new, coupled with an increasingly mobile human population has created many novel opportunities for potential pathogens to meet new human hosts. All of this requires new prevention and control strategies but progress has been slow, despite recent technological advances and increased investment. The rapid increase in data proved difficult to translate into practical applications for human health care, and new approaches and analyses are needed to make the most of new opportunities. One of these is the use of the new tools of systems biology and this chapter will review the application of these to microbial pathogens.

Host-microbe interactions are complex phenomena spanning multiple levels of complexity, from environmental and ecological factors up to the cellular and genetic levels of host responses. At each of these levels a relationship is established between one or more microorganisms and the host, resulting in formation of various forms of associations ranging from symbiosis to parasitism.

Pathogens have the potential to cause disease in their hosts through host-pathogen interactions in which host defences are challenged by the invasive capacities of the pathogen. The aim of this chapter is to give an outline of attempts made to unravel the components of host-pathogen interactions at the cellular and molecular levels and to discuss strategies to skew the balance in favour of the host, focusing our attention on crucial intracellular pathogens causing globally relevant diseases such as tuberculosis, gastroenteritis, influenza and malaria.

Phagocytosis is the fundamental cellular process by which eukaryotic cells bind and engulf particles by deforming their plasma membrane. Particle engulfment involves particle recognition by cell-surface receptors, signalling, and remodelling of the actin cytoskeleton to guide the membrane around the particle in a zipper-like fashion. The signalling complexity is daunting, involving hundreds of different molecular species during the initial stages of engulfment. For instance, the well-characterised immune Fcã and the integrin CR3 receptors signal to tyrosine kinases and Rho GTPases, ultimately regulating a wide variety of proteins, which direct actin polymerisation and myosin-motor proteins for force generation and contraction.

Despite the signalling complexity, phagocytosis also depends strongly on simple biophysical parameters, such as shape and cell stiffness, or membrane biophysical properties that are independent of the type of cell or particle. We argue that these emergent, universal features are particularly important to address in order to explain this evolutionary well-conserved and robust mechanism. In this chapter we review these universal features to describe the principles of phagocytosis. Specifically, we use a recently published model of phagocytic engulfment as a guide. Finally, we discuss state-of-the-art live-cell fluorescence microscopy, recently used to elucidate the dynamics of phospholipids, actin polymerisation and myosins in the particle-shape recognition by the amoeba *Dictyostelium*.

Despite decades of research many aspects of the biology of *Mycobacterium tuberculosis* remain unclear and this is reflected in the antiquated tools available to treat and prevent tuberculosis. Consequently, this disease remains a serious public health problem responsible for 2 to 3 million deaths each year.

Important discoveries linking *M. tuberculosis* metabolism and pathogenesis have renewed interest in the metabolic underpinning of the interaction between the pathogen and its host. Whereas, previous experimental studies tended to focus on the role of single genes, antigens or enzymes the central paradigm of systems biology is that the role of any gene cannot be determined in isolation from its context.

Therefore systems approaches examine the role of genes and proteins embedded within a network of interactions. We here examine the application of this approach to studying metabolism of *M. tuberculosis*.

Recent advances in high throughput experimental technologies, such as functional genomics and metabolomics, provide datasets that can be analysed with computational tools such as flux balance analysis.

These new approaches allow metabolism to be studied on a genome scale and have already been applied to gain insights into the metabolic pathways utilised by *M. tuberculosis in vitro* and identify potential drug targets. The information from these studies will fundamentally change our approach to tuberculosis research and lead to new targets for therapeutic drugs and vaccines.

Life at high temperature challenges the stability of macromolecules and cellular components, but also the stability of metabolites, which has received little attention. For the cell, the thermal instability of metabolites means it has to deal with the loss of free energy and carbon, or in more extremes, it might result in the accumulation of dead-end compounds.

In order to elucidate the requirements and principles of metabolism at high temperature, we used a comparative blueprint modelling approach of the lower part of the glycolysis cycle. The conversion of glyceraldehyde 3-phosphate to pyruvate from the thermoacidophilic Crenarchaeon *Sulfolobus solfataricus* P2 (optimal growth-temperature

80°C) was modelled based on the available blueprint model of the eukaryotic model organism *Saccharomyces cerevisiae* (optimal growth-temperature of 30°C).

In *S. solfataricus* only one reaction is different, namely glyceraldehyde-3-phosphate is directly converted into 3-phosphoglycerate by the non-phosphorylating glyceraldehyde-3-phosphate dehydrogenase, omitting the extremely heat-instable 1,3-bisphosphoglycerate. By taking the temperature dependent non-enzymatic (spontaneous) degradation of 1,3-bisphosphoglycerate in account, modelling reveals that a hot lifestyle requires a cool design.

2

Diversity of Abyssal Marine Life

The Census of the Diversity of Abyssal Marine Life (CeDAMar) was devoted to the study of the largest and remotest ecosystem on Earth, the major deep basins stretching between continental margins and the mid-ocean ridge system. Abyssal plains and basins account for about half of Earth's surface (Tyler 2003) and harbour a great variety of life forms. As part of the overall Census of Marine Life, the field project CeDAMar was designed to study the diversity, distribution, and abundance of organisms living in, on, or directly above the seafloor. Prominent features such as ridges, seamounts, trenches, and chemosynthetic environments were covered by other Census projects.

Abyssal Plains

Until the late nineteenth century, abyssal sediments were believed to be azoic deserts owing to a lack of sunlight and primary production. This view changed dramatically with the British Challenger expedition (1872–1876), which found deep-sea life throughout the world ocean. Modern marine diversity research began in the 1960s when Sanders, Hessler, and co-workers were able to show that the abundance of macrobenthic organisms decreased with depth whereas the number of species increased (Sanders et al. 1965; Hessler & Sanders 1967; Sanders & Hessler 1969). Pivotal in the development of the scientific interest in marine diversity patterns was a study by Grassle & Maciolek (1992) of a series of box corer samples collected along a 176 km transect on the northwest Atlantic continental slope.

Species turnover rates along the transect suggested that the number of species at the deep-ocean floor may rival that of tropical rainforests. This study led to broad debate about the number of marine species and the distribution of diversity along bathymetric and latitudinal gradients (Poore & Wilson 1993; Rex et al. 1993, 1997; Thomas & Gooday 1996; Culver & Buzas 2000). Before the year 2000, biological research in the abyss had been conducted only sporadically as part of the classic worldwide expeditions aboard American, German, Danish, and Swedish vessels around the turn of the century into the mid 1990s. More recently, between 1948 and 2000, the P.P. Shirshov Institute sampled more than 1,700 stations below 3,000 m including abyssal plains, basins, and trenches down to 9,000 m. Studies of abyssal diversity and biogeography were complicated by the logistic challenges of deep-sea exploration. When the first CeDAMar expeditions were planned, the total sampled area of deep-sea floor was equal to no more than a few football fields, and by the year 2005 the total sampled area below 4,000 m amounted to about 1.4×10^{-9}% (Stuart et al. 2008).

The CeDAMar Rationale

When CeDAMar was initiated, published results suggested that deep-sea sediments supported low biotic abundance and biomass, but potentially high species richness depending on taxon. All expeditions to abyssal plains and basins showed that regardless of the location, roughly 90% of the infaunal species collected in a typical abyssal sample were new to science.

Open Questions in Deep-Sea Research

One fundamental gap in our knowledge of the abyss was the existence of vast geographic areas that had not been sampled, for example, the central Pacific Ocean and oceans of the southern hemisphere, because they were so remote from oceanographic institutions. CeDAMar expeditions were specifically designed to explore both sides of the southern Atlantic, southern Indian Ocean, and the Southern Ocean; the Northeast Atlantic; the central Pacific; and, as an example for a warm, ultra-oligotrophic deep sea, the eastern

he occurrence of high biodiversity in the extreme habitat conditions that characterise the abyss, such as low temperature, very high hydrostatic pressure, little habitat complexity, and extremely low food availability, was perceived to be one of the major biogeographic puzzles of our time. Despite the potential importance of this vast ecosystem as a reservoir for genetic diversity and evolutionary novelty, very little was known about the factors regulating deep-sea species richness (Gage & Tyler 1991; Gray 2002). CeDAMar therefore aimed to collect new reliable data on species assemblages of ocean basins and determine the large-scale distribution of species among these basins. Documentation of the actual species diversity of abyssal plains provided a baseline for global-change research and for a better understanding of historical causes and ecological factors regulating biodiversity.

Even less is known about the biology of abyssal organisms. One of the unanswered questions in this context was the relation between food supply and the number of species present in a given deep-sea area. The deep-sea benthos depends ultimately on surface production that sinks through the water column. Although it seems evident that the biomass of deep-sea organisms should be positively correlated with food availability (Rowe 1971; C.R. Smith et al. 1997; Brown 2001), the productivity–biodiversity relationship is less clear.

Specific CeDAMar Questions

Considering our lack of knowledge, CeDAMar focused research efforts in a way that would produce tangible results within a set timeframe of less than ten years. Deep-sea biologists identified the most urgent questions to be addressed by CeDAMar expeditions, keeping in mind the overarching Census themes of diversity, abundance, and distribution.

Questions Concerning Diversity

- How does diversity vary at different geographic scales, between different size classes of organisms, and with differences in food supply?
- Are there centres of high diversity (hot spots of diversity) in the deep sea?
- What is the role of evolutionary-historic processes in determining diversity levels?
- How do manganese nodules or drop stones influence benthic diversity?

Questions Concerning Abundance

- How do organisms of different size classes respond to environmental factors?
- What is the relation between food availability and benthic standing stock?

Questions Concerning Distribution

- Do biogeographic barriers affect the distribution of abyssal fauna? How endemic is the abyssal fauna?
- How common are cosmopolitan species in the abyss? Is there gene flow between distant abyssal communities of the same species?

- Are there latitudinal gradients in species richness? Is the diversity of a given basin similar to the diversity of basins in other oceans at similar latitudes.

Finding Answers: Methods and Programs of CeDAMar

The most prominent reason why the abyss has been explored to such a small degree is the difficulty of reaching it. Apart from the scarcity of research vessels, there are many logistic challenges, the time required for sampling great ocean depths not being the least. To lower sampling gear to the seafloor some 4,500 m below the surface and retrieve it back to the ship, several hours are necessary for each single sampling. The control of the actual sampling process on the bottom is limited by the great depth and the amount of wire between ship and gear.

The methodology that CeDAMar used was more traditional than hitech, consisting of coring devices (box corer and multicorer), epibenthic sledges, Agassiz trawls, and, when possible, a sediment profiling camera with or without a video camera. This set of gear was used in a standardised way to ensure (1) collection of organisms of all size classes from bacteria to large epifauna such as corals, sea anemones, sponges, holothurians, and stalked crinoids, and (2) comparability of results among CeDAMar projects and with the existing literature. The Time Series study of the seafloor in the Porcupine Abyssal Plain used a time-lapse camera and sediment traps to monitor processes on the seafloor.

Project DIVA

DIVA (diversity gradients in the Atlantic) is the seed project of CeDAMar, with the main focus on the question of latitudinal gradients in biodiversity in the southern Atlantic. Sampling locations were the abyssal basins off west Africa from the Cape to the equator and the Argentine and Brazil basins off the east coast of South America.

Project ANDEEP

ANDEEP (Antarctic benthic deep-sea biodiversity – colonisation and recent community patterns) was dedicated to the abyssal waters in the Atlantic sector of the Southern Ocean. This region is one of the least investigated and it closed the gap between the two study areas of DIVA. It is also the location closest to the pole and farthest away from the equator, which made it very suitable to prove or disprove that a decline in marine biodiversity is present from the equator to the poles.

Projects KAPLAN and NODINAUT

The study area of KAPLAN and NODINAUT was the manganese nodule field in the Clarion-Clipperton Fracture Zone (CCZ), with the main focus centred on the question of the impact of nodules on biodiversity at different scales. Results were used for recommendations concerning marine protected areas (MPAs) to protect the fauna in case of nodule mining. In light of increasing demand for minerals, deep-sea mining has become a realistic possibility

Project Biozaire

Biozaire was conducted off West Africa, just inshore of the DIVA area, encompassing the deep slope, abyssal plain, and a chemosynthetic site (a so-called pockmark). The objective was to characterise the "benthic community structure in relation with physical and chemical processes in a region of oil and gas interest" (Sibuet & Vangriesheim 2009).

Project LEVAR

LEVAR (Levantine Basin Biodiversity Variability) was one of the younger projects of CeDAMar, the study area being the eastern Mediterranean Sea with its comparatively shallow abyss (around 3,000 m), warm water at depth, and extremely poor food supply. Stations near Crete were sampled during one cruise. The aim was to determine whether proximity to shore or the depth was more important in influencing community composition and the distributions of abyssal biota.

Project CROZEX

The relation between surface primary production and benthic community composition was also explored during three cruises of the CROZEX (Crozet circulation iron fertilization and export production experiment) expedition off the sub-Antarctic Crozet Islands (Indian Ocean). The background of this study was a proposal put forward by biogeochemists suggesting that natural iron fertilization might enhance algal growth, which would sink to the abyssal seafloor, thus sequestering CO_2 and taking it out of the atmosphere. By observing processes driven by natural fertilization through iron eroded from the islands, CROZEX was designed to assess whether artificial iron fertilization might be a feasible option to fight global warming.

Project Time Series

A time-lapse camera system and moorings including sediment traps have been used to observe the deep ocean floor in the Porcupine

Abyssal Plain since 1989, changing our perception of the quiescent, stable abyss to that of a very dynamic environment with sometimes radical changes in communities. One incident, the so-called *Amperima* Event named after the sea cucumber *Amperima rosea*, has become famous because of substantial changes in abundance related to changes in food supply

Project ENAB

Evolution in the deep sea was the focus of ENAB (Evolution in the North Atlantic Basin), with a sampling cruise conducted along the famous Gay Head–Bermuda transect that in the early 1960s had started biodiversity research in the deep sea. The program was dedicated to assessing spatial population genetic structure in deep-sea mollusks to determine patterns of population differentiation, speciation, and phyletic evolution.

CeDAMar Database

One of the legacies that may prove to be highly valuable to deep-sea researchers today and in the future is a freely accessible database that will be maintained and updated beyond the life of CeDAMar. So far, some 12,000 records, representing more than 3,000 species from nearly 4,800 locations distributed in all oceans can be queried. These records are made available to Ocean Biogeographic Information System, the database of the Census, from where they can also be accessed by anyone. With a special tool, maps can be created with different resolutions. This section shows the number of abyssal records per area, in this case a grid of 10 degree × 10 degree squares (roughly 100 km × 100 km).

There are four areas with relatively extensive sampling on which much of our knowledge of the abyssal fauna is based: (1) the northwest Atlantic off the US east coast sampled in the 1980s, including stations on the continental slope that led to the estimates of deep-sea species richness by Grassle & Maciolek (1992); (2) the manganese nodule area off Peru, where the German DISCOL disturbance experiment was performed in the 1980s and 1990s to assess recovery of abyssal benthic fauna after massive disturbance mimicking possible effects of nodule mining; (3) the Porcupine Abyssal Plain and Gulf of Gascogne where British and French deep-sea investigations were concentrated; and (4) the Kurile–Kamchatka Trench, which was a main study area of Russian deep-sea research.

The remaining area of the abyssal plains is still unsampled or poorly sampled, showing that even the substantial effort put into abyssal expeditions during CeDAMar has relatively little effect on sample coverage from a worldwide perspective.

Overcoming the Taxonomic Impediment

As all knowledge about ecosystems is based on knowing the identity of species in a particular system, much effort has been put into overcoming the so-called taxonomic impediment. The term means the general lack of specialists for identification of marine animals. Workshops and short-term stays at participating institutions (taxonomic exchanges) have helped to foster communication and intercalibration of the numerous personal databases from a broad range of projects. For polychaetes, a platform was created with the help of the Natural History Museum (London) to exchange information by the Internet on yet unpublished but already well-defined "working species", allowing specialists to share information on an additional 50–90% of their respective taxa. A more visible outcome for the entire scientific community was CeDAMar's goal to deliver formal descriptions of 500 new abyssal species by the end of the first Census in October 2010. The goal will have been reached by the time this book is published. Nearly half of all newly described or redescribed species are crustaceans (243 species, 91 of which are isopods), followed by nematodes (55 species) and mollusks (41 species, including 32 gastropods).

Major Results

Through the results generated by the CeDAMar project our perception of the abyss has changed fundamentally. This change in perception may be condensed into two statements which, although they may seem trivial at first glance, are significant changes in how scientists view the abyss: (1) extreme is normal; (2) rare is common.

Extreme Is Normal

Quite surprisingly, scientists even in the twentieth century viewed remote habitats on Earth from an anthropocentric perspective. The richness of life on abyssal seafloors showed quite convincingly that this habitat, which is extreme or even "inhospitable" to us, is highly habitable for a remarkable range of organisms. Even though we still know very little about the biology of abyssal organisms, it has become very apparent that many are well adapted to "extreme" conditions; reproduction takes place as well as speciation, and observations of a single site over time, such as the Porcupine Abyssal Plain (PAP) Time Series project, revealed that the abyssal seafloor can be unexpectedly dynamic. The massive bloom of the holothurian *Amperima rosea* in the PAP observed in the late 1990s was followed by a significant shift in the communities of several other deep-sea invertebrates that was documented over a period of 20 years (Billett et al. 2009). Not all other organisms seem to be affected by the alterations of the environment. Some of the polychaete populations, for example, did not react in any visible way, whereas others showed a significant increase in the number of individuals which could be related to increased nutrient input.

Rare Is Common

In terms of the general structure of benthic communities, there are large differences between the abyss and shallower environments. Nearly all species found in the abyss are rare, at least to our current knowledge. In practical terms it means that most species have been recorded as one or two individuals from one or two sampling sites, even in large programs during which thousands of animals were collected. With very few exceptions, none of the communities sampled during CeDAMar expeditions were characterised by one or a few numerically dominating species as is typically the case in shelf communities.

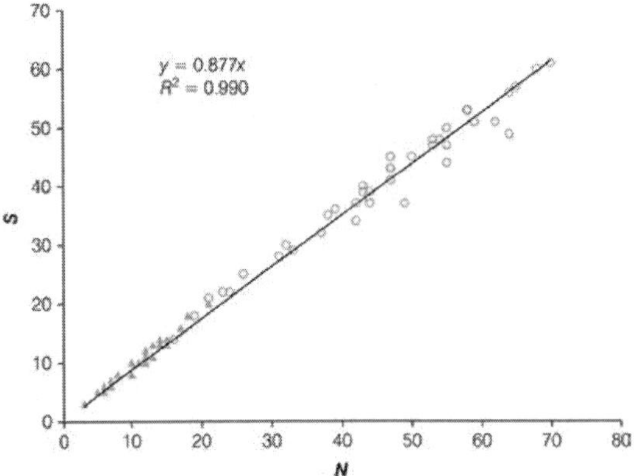

Figure: *Copepod species numbers (S) and corresponding numbers of specimens (N) collected by multicorer at two stations in the Angola Basin (triangles, station 325; circles, station 346) with line of linear regression. The number of individuals nearly equals the number of species. From Rose et al. 2005.*

Diversity of Abyssal Benthos

One of the ways to measure diversity is to look at the number of species at one particular site (alpha diversity), in addition species turnover along a certain distance (beta diversity) may also be assessed.

Both measures of diversity were found to be much higher than expected. For example, copepods in the southeast Atlantic occurred everywhere in high abundances, but most species were undescribed (DIVA cruises): 98% of these species had never been seen before. Even smaller animals, the unicellular foraminiferans, showed high species turnover rates in the manganese nodule fields in the Pacific. At sampling sites no more than roughly 600 miles apart, different communities of foraminiferans were found. However, not all foraminiferan distributions appear to be restricted. In another study, including the ANDEEP material, other foraminiferans were discovered that are distributed from pole to pole, obviously coping with many very different habitat conditions.

Habitat heterogeneity is considered to be one of the major drivers of biodiversity because it provides a greater range of niches for the formation of new species. The abyssal seafloor was found to be as heterogeneous as shallower areas, perhaps most obviously in manganese nodule fields of the equatorial Pacific and in the Southern Ocean where stones drop out of melting icebergs and provide greater heterogeneity in substrata.

The community structure of abyssal megafauna and macrofauna in manganese nodule fields was found to differ not only due to the availability and quality of food but also because of the heterogeneity in physical and chemical properties of the habitat (nodules and superficial sediment). Studies undertaken at the local scale (1–5 km in distance) with the manned submersible Nautile showed for the first time that nodule fields constitute a distinct habitat for infaunal communities, and that macrofauna and meiofauna components differ in abundance depending on the presence of nodules (Miljutina et al. 2009).

The geologic history of a basin can play an important role for biodiversity as well. A good example is the Southern Ocean. Its history includes not only periods of anoxia in the late Jurassic and cooling in the late Eocene/early Oligocene, but also cycles of glaciation and deglaciation which led to migration of shallow-water species into bathyal and abyssal depths (submergence) as well as recolonisation of shallow sea bottoms from the deep (emergence). Applying molecular methods, Raupach et al. (2004, 2009) showed that shallow-water isopods colonised the deep sea at least on four separate occasions. Several isopod families underwent spectacular radiation events in the abyss, resulting in an exceptionally high number of species and species complexes. The Scotia and Weddell Seas, the geographic focus of the ANDEEP investigations, are characterised by a complex tectonic history related to the Middle Jurassic break-up of the Gondwana supercontinent which began around 180 million years (Ma) ago (Storey 1995). The Scotia Sea is much younger and formed during the past approximately 40 Ma (Thomson 2004). However, it is unknown whether the great biodiversity documented for many taxa in the deep Weddell Sea can be explained by the age of the ocean floor.

Another example is the generally low diversity of the benthos in the deep Mediterranean Sea, which is related to, among other reasons, the complex paleoecological history characterised by the Messinian salinity crisis and the almost complete desiccation of the basin.

Spatial and Temporal Variability in Primary Productivity in the World's Oceans and Its Effects on Abyssal Communities

Changes in primary productivity in the surface waters of the world's oceans are mirrored in abyssal communities in both space and time (C.R. Smith et al. 2008a). Organic matter created by photosynthetic production provides the food for most deep-sea life. Changes in food production at the sea surface, therefore, and the subsequent transport of organic matter into the ocean's interior through the biological carbon pump, have a profound effect on life on the abyssal seafloor.

It is well known that in regions where seasons are evident in surface waters, seasonal changes occur on the deep-sea floor within a matter of weeks (Billett et al. 1983; C.R. Smith et al. 1997; Beaulieu 2002). Large-scale biogeographical provinces in surface waters are reflected in broad changes in the structure of abyssal communities (Smith C.R. et al. 2008a). Decadal-scale shifts in primary production, caused by climate-related oscillations, produce long-term radical changes in deep-sea communities (Billett et al. 2001, 2009; Ruhl & Smith 2004; Ruhl 2007; C.R. Smith et al. 2008a; Smith K.L. et al. 2009). The fall of the carcasses of whales and fish (C.R. Smith & Baco 2003) and the mass deposition of jellyfish (Billett et al. 2006) provide additional, if localised, organic inputs. The abyss is linked intimately to processes at the sea surface.

CeDAMar projects have contributed significantly to recent advances made in our understanding of how surface water productivity affects abyssal ecosystems. Spatial variations in the distribution of species have been related to changes in surface water productivity in the Kaplan, DIVA, and CROZEX projects. In addition, radical changes in abyssal communities with time have been documented at the PAP in the Northeast Atlantic Ocean. Similar large-scale changes with time have been noted in the northeast Pacific Ocean (K.L. Smith et al. 2009).

At the PAP, CeDAMar has documented how over a 20 year time series (1989 to 2009) the abyssal megafauna changed in total abundance by two orders of magnitude in 1996 (Billett et al. 2009). This was mainly due to the increase in the holothurian species *Amperima rosea* and became known as the "*Amperima* Event" (Billett et al. 2001). Significant changes in the abundances of several megafaunal taxa occurred, including ophiuroids, actiniarians, pycnogonids, tunicates, and holothurians other than *A. rosea*. The changes were evident over a vast area of the abyssal plain (Billett et al. 2001) and had a significant effect on the recycling of organic matter at the sediment surface (Bett et al. 2001). During the CeDAMar project it has been determined that protozoan and metazoan meiofauna (Gooday et al. 2010; Kalogeropoulou et al. 2010) and polychaete macrofauna (Soto et al. 2009) also increased significantly in abundance during the "*Amperima* Event". All elements of the benthic community showed a simultaneous change indicative of a large environmental event.

Protozoan phytodetritus indicator species showed a sharp decrease in abundance, whereas trochamminaceans, which previously had been comparatively rare, became dominant, potentially because of the increased disturbance caused by the megafauna (Gooday et al. 2009). In the metazoan meiofauna increases in abundance were seen in the nematode and the meiofaunal polychaetes, but not in the copepods. Ostracods decreased in abundance. The three dominant macrofaunal polychaete families, Cirratulidae, Spionidae, and Opheliidae, all increased in abundance but no major changes occurred in the community structure and dominant species (Soto et al. 2009), unlike the megafauna.

These results show that abyssal benthic communities change significantly with time. Similar results in the northeast Pacific Ocean indicate that such phenomena are widespread in productive regions of the world's oceans (K.L. Smith et al. 2009). The flux of organic matter may change by about an order of magnitude from one year to the next (Lampitt et al. 2010) and abundances in fauna have been shown to be correlated to climate indices that influence the biological carbon pump on regional scales (K.L. Smith et al. 2006, 2009).

Although many elements of the benthic community change at the same time in the Time Series studies, the scale of the response is not the same in all taxa or size classes. Larger changes in abundance are apparent in the megafauna and there are greater changes in the dominant species. This has important implications for interpreting geographic variations in the distributions of species in the different size classes of the benthic community.

Annual particulate organic carbon (POC) flux and benthic parameters have been measured together at only a few sites in the abyssal ocean. However, where POC flux has been measured directly, there are strong linear relations between POC flux and the abundance and/or biomass of specific biotic size classes, including megafauna, macrofauna, and microbes (C.R. Smith et al. 1997; C.R. Smith et al. 2008a; K.L. Smith et al. 2009). Average biomass of megafauna (Lampitt et al. 1986) and macrofauna (Rowe 1971) decline significantly with increasing water depth (and hence decreasing POC flux), resulting in the smaller size classes (bacteria and meiofauna) dominating community biomass at abyssal water depths (greater than 3,000 m) (Rex et al. 2006).

Despite this, experimental results (Witte et al. 2003) and time-lapse photography (Bett et al. 2001) indicate that larger organisms play important functional roles in energy flow through food-limited abyssal ecosystems by outcompeting the smaller size classes for freshly deposited detritus. Changes in the spatial distribution of abyssal fauna therefore not only reflect the total input of organic matter, but also the periodicity and predictability in its supply. In addition, changes may be related to the quality of the organic matter (Ginger et al. 2001; Wigham et al. 2003; FitzGeorge-Balfour et al. 2010).

In another CeDAMar study around the Crozet Islands in the southern Indian Ocean, the distributions of protozoan and metazoan meiofauna, and of megafauna, were studied in relation to an area of natural iron fertilization in the oceans (Pollard et al. 2009). Iron carried off the volcanic islands of Crozet leads to seasonal phytoplankton blooms to the north of the Crozet plateau, as opposed to the south of the islands where iron is limiting. The eutrophic site had a greater diversity of live foraminiferans, and the phytodetritus indicator species *Epistominella exigua* was more abundant at this locality (Hughes et al. 2007).

In contrast, the megafaunal communities in the two areas were radically different (Wolff et al. personal communication). The most abundant species *Peniagone crozeti* (Cross et al. 2009), found only at the seasonally productive site, was new to science. This indicates that megafaunal communities may be the most sensitive to changes in surface water productivity, whereas the smaller size fractions may show broader distributions, depending on the taxon. However, broad generalisations are difficult to make because certain macrofaunal species, including isopods and polychaetes, are restricted to productive areas of the ocean, such as the Southern Ocean (Brandt et al. 2007a, 2007b, 2007c).

Latitudinal/Depth Gradients of Biodiversity in the Atlantic Ocean

Latitudinal gradients are the most conspicuous and ubiquitous biogeographic patterns in terrestrial and coastal ecosystems, but their explanation remains elusive. They were long assumed not to occur in the deep sea because the deep overlying water column buffered communities from the climatic phenomena thought to ultimately shape large-scale patterns of diversity. However, there is evidence that latitudinal gradients of diversity do exist in several macrofaunal taxa and foraminiferans in bathyal communities. They have not been examined previously at abyssal depths, largely because there are so few abyssal samples. The comprehensive DIVA datasets are being used to test whether latitudinal gradients do exist at abyssal depths. The results will be especially interesting because it is unclear whether latitudinal gradients in macrofaunal taxa exist in the southern hemisphere (Rex et al. 2000).

Results from the ANDEEP expeditions have shown that the impact of depth on species richness is not consistent among taxonomic groups. Ellingsen et al. (2007) examined general macrofaunal response to water depth in the Atlantic sector of the deep Southern Ocean using data on polychaetes, isopods, and bivalves collected during the EASIZ II (Ecology of the Antarctic Sea-Ice Zone, 1998) and ANDEEP I and II cruises (2002), ranging from 774 to 6,348 m depth. They found that the isopods displayed higher species richness in the middle depth range (216 species in 3,000 m depth) and lower in the shallower and deeper parts of the area (Brandt et al. 2005), as reported for other deep-sea areas.

However, the number of bivalve species showed no clear relation to depth, and polychaetes showed a negative relation to depth (Ellingsen et al. 2007). Although the data were collected over a wide geographical area (58°14¢–74°36¢ S, 22°08¢–60°44¢ W), the numbers of isopod, polychaete, and bivalve species did not show any consistent relation to latitude or longitude. Gastropods and bivalves show a variety of diversity–depth patterns among deep-sea basins (Allen 2008; Stuart & Rex 2009). Brandt et al. (2009) investigated the bathymetric distribution patterns of bivalves, gastropods, isopods and polychaetes in the Southern Ocean from 0 to 5,000 m, and found that the patterns differed between the different taxonomic groups.

Diversity and Biogeography of Antarctic Deep-Sea Fauna

Within the Southern Ocean, the abyssal benthic realm is the largest ecosystem and covers 27.9 million km^2 (Clarke & Johnston 2003). The Southern Ocean is characterised by some unique environmental features, which include a very deep continental shelf and a weakly stratified water column. It is also the source for the deep-water production influencing the deep circulation throughout the world. These physical characteristics led to the assumption that the Southern Ocean deep-sea fauna may be related both to adjacent shelf communities and to those living in other deep oceans. In the past century, Antarctic benthic shelf communities have been investigated extensively and are known to be characterised by high levels of endemism, gigantism, slow growth, longevity, and late maturity.

Some amphipod, isopod, and fish families have adaptive radiations which have led to considerable novel biodiversity in these groups. Contrary to the Southern Ocean shelf, little was known about life in the vast Southern Ocean deep-sea region before the ANDEEP project. ANDEEP was a multidisciplinary international project which involved two expeditions to the Weddell and Scotia Seas in 2002 (Brandt & Hilbig 2004) and a third expedition (ANDEEP III) in 2005 to the Cape and Agulhas Basins, Weddell Sea, Bellingshausen Sea, and Drake Passage.

In total, 40 stations were sampled between 748 and 6,348 m water depth with a focus on the abyss (Brandt & Hilbig 2004; Brandt & Ebbe 2007; Brandt et al. 2007a, 2007b, 2007c). The analyses revealed an astonishingly high biodiversity of several different taxa. From the material analysed, more than 1,400 species were identified, and of these, more than 700 were new to science. In some groups of organisms, such as nematodes and isopods, greater than 90% of the species collected were new to science. Among the most important isopod families, over 95% of the species collected were unknown (Brandt et al. 2007a; Malyutina & Brandt 2007). Although we know that some species complexes have radiated in the deep Southern Ocean (Brökeland & Raupach 2008; Raupach & Wägele 2006; Raupach et al. 2007), it is unclear whether they have evolved here and subsequently spread into other ocean basins. Many species (>50%) were rare or patchy and occurred at only one station. Many species were singletons.

Biogeographic and bathymetric trends varied between groups and were probably related to differences in the reproductive mode (Brandt et al. 2007b, 2009; Pearse et al. 2009). In the isopods and polychaetes, slope assemblages included species that have invaded from either the shelf or the abyss through emergence or submergence, respectively, whereas in other taxa such as bivalves and gastropods, the shelf and slope assemblages were more distinct. Abyssal faunas tended to have stronger biogeographic links to other oceans, particularly the Atlantic, but mainly for organisms with good dispersal capabilities such as the foraminiferans (Brandt et al. 2007b; Pawlowski et al. 2007) and polychaetes (Schüller & Ebbe 2007; Schüller et al. 2009).

The isopods, ostracods, and nematodes, which are poor dispersers, include many species currently known only from the Southern Ocean. In some groups, such as the Munnopsidae (Isopoda), the highest number of species (219) was reported in a worldwide biogeographical analysis (Malyutina & Brandt 2007). The ANDEEP results challenge the hypothesis that deep-sea diversity is depressed in the Southern Ocean and provide a sound basis for future explorations of the evolutionary significance of the varied biogeographic patterns observed in this remote environment.

The Mediterranean Sea: Diversity Patterns in a Warm Deep Sea

The Mediterranean region is characterised by the presence of both low and very high biodiversity, high levels of endemism are apparent, and in some areas strong energetic gradients in primary production and food supply to the deep occur decreasing from the western to the eastern basins and from shallower to deeper sites. The deep Mediterranean has generally been considered to have lower diversity than other deep-sea regions.

Faunal exchange with the Atlantic Ocean is impaired by differences in deep-sea temperatures (approximately 10 °C higher in the Mediterranean than in the Atlantic Ocean at the same depth), which makes the establishment of incoming deep Atlantic fauna difficult. In particular, the abyssal basins of the Eastern Mediterranean are extremely unusual deep-sea systems with water temperatures at 4,000 m in excess of 14 °C. Barriers to colonisation from the Atlantic also include salinity gradients and differences in food supply, as well as the existence of shallow sills. The deep Mediterranean is thus generally considered a "biological desert", although certain areas display such high benthic activity as to be characterised as "benthic hot spots". These areas are in most cases located at or near the mouth of submarine canyons that transport, through flash flooding, sediment failures, and dense shelfwater cascading, large amounts of sediment and organic material to the deep-sea floor (Canals et al. 2006). Abyssal trenches act as traps of organic matter of either terrestrial or pelagic origin (Tselepides & Lampadariou 2004; Boetius et al. 1996). Large-scale hydrographic changes (Eastern Mediterranean Transient) have also been implicated in enhancing the productivity of the euphotic zone and indirectly structuring the underlying deep benthic communities (Danovaro et al. 2004).

The Mediterranean differs from other deep-sea ecosystems in terms of its megafaunal species composition (Jones et al. 2003). Typical deep-water groups, such as echinoderms, glass sponges, and macroscopic Foraminifera (Xenophyophora) are absent in the deep Mediterranean, whereas other faunistic groups (fishes, decapod crustaceans, mysids, and gastropods) are represented poorly compared with the Northeast Atlantic.

Although the low-diversity pattern is based on the analysis of macro- and megabenthos, recent evidence (Danovaro et al. 2008) suggests that the Mediterranean deep-sea nematode fauna is rather diverse and cannot be considered "biodiversity depleted". In fact, it was suggested that meiofaunal benthic biodiversity in the deep Atlantic and Mediterranean basins is similar.

A detailed analysis of food availability in the deep Mediterranean revealed that organic matter composition differed between the east and the west Mediterranean. Organic matter in the east was dominated by a high fraction of proteins and lipids. Therefore, although there were reduced amounts of organic matter in the east, this was to a certain extent compensated for by higher food quality and bioavailability. It seems that biodiversity patterns are not controlled by the amounts of food resources alone but also by the availability of the organic matter.

The project LEVAR explored not only the composition of benthic communities, but also environmental factors such as distance from shore, that is, supply of nutrients from shallower areas nearby, versus primary production in surface waters right above the sampling site and their respective influence on diversity. Preliminary results show that the benthic fauna at abyssal sites of the eastern Mediterranean is extremely poor in terms of abundance during normal oligotrophic periods, but can quickly develop high biomass when pulses of organic material settle down to the seafloor after unpredictable phytoplankton

Abyssal Diversity Hot Spots

The diversity of life in the Southern Ocean (Brandt & Hilbig 2004) and the central Pacific Ocean (Glover et al. 2002) is high enough to characterise these areas as abyssal biodiversity hot spots. Glover et al. (2002) stated, "Local polychaete species diversity beneath the equatorial Pacific upwelling (measured by rarefaction) appears to be unusually high for the deep sea, exceeding by at least 10 to 20% that measured in abyssal sites in the Atlantic and Pacific, and on the continental slopes of the North Atlantic, North Pacific, and Indian Oceans." The use of molecular genetic methods will likely reveal an even higher diversity as many organisms looking alike under the microscope turn out to belong to different species, discernible only by differences in their genes.

Abundance of Abyssal Benthos

Studies of the CeDAMar project Biozaire on the continental slope of the Gulf of Guinea, adjacent to the DIVA 1 study area in the abyss, revealed that benthic communities living closer to shore are influenced by a very complex system of environmental parameters. Nevertheless, as in the PAP, the megafauna seemed to respond most directly to the influence of the organic material supplied by the Congo channel, whereas densities of smaller organisms macrofauna and meiofauna were subject to changes in environmental parameters, particularly in trophic inputs, at regional scale beyond the effects of the Congo channel (Sibuet & Vangriesheim 2009). Two of three study sites were located in approximately 4,000 m depth, 15 and 150 km south of the Congo channel. They were sampled during three cruises that were roughly one and two years apart.

Abundance of macro and meiofauna increased substantially between 2001 and 2003, but interestingly not near the Congo river fan where increased input of organic matter was observed but rather at the station away from the fan. Obviously it was the quality of the food rather than the quantity that had the most profound effect on abundance, the organic matter near the Congo channel being mostly terrigeneous and, thus, of lower value for the deep benthos. These results agree well with findings from the deep eastern Mediterranean Sea.

Distributional Patterns in the Abyss: Endemism Versus Cosmopolitanism

The traditional view of an abyssal cosmopolitan fauna has been strongly favoured given the enormous, contiguous nature of abyssal environments, and the isolated records of apparently conspecific animals in separate ocean basins. Recent CeDAMar field projects such as the ANDEEP cruises in the Southern Ocean, the KAPLAN cruises in the central Pacific, and the DIVA cruises in the south Atlantic have created new opportunities to re-assess degrees of cosmopolitanism, which are reviewed here.

CeDAMar scientists have focused on a range of dominant abyssal taxa, which exhibit a range of reproductive strategies. These include peracarid crustaceans, copepods, polychaete worms, mollusks, holothurians, and foraminiferans. Peracarids generally brood young in their marsupium and there is no distinct larval stage (Brandt et al. personal communication). Copepods are direct-developing, with juveniles and adults both probably distributed by ocean currents. Polychaetes include species that either brood or display a biphasic life cycle with free-swimming planktotrophic or lecithotrophic larvae: both modes are thought to occur in abyssal species (Beesley et al. 2000). Deep-sea gastropods and bivalves generally reproduce by planktotrophic or lecithotrophic larval dispersal (Rex et al. 2005). Deep-sea holothurians have a broad range of egg sizes, from 180 to 4,000 μm (Billett 1991). The largest egg sizes are thought to lead to direct development of free-swimming juvenile holothurians within the abyssopelagic zone allowing for wide dispersal (Billett et al. 1985). Abyssal foraminiferans are thought to reproduce asexually (Murray 1991).

A study of cosmopolitanism in 45 deep-sea peracarid species has revealed only 11 species which occur in all oceans studied (the North Atlantic, South Atlantic, Southern Ocean, North and South Pacific, and Indian Oceans) (Brandt et al. personal communication). However, 33 species have distributions across more than one ocean basin, and 16 species are shared between the North Atlantic and North Pacific. Molecular-based studies of asellote isopods have revealed cryptic species, but these studies have so far been limited to a small range of taxa (Raupach et al. 2009). For benthic harpacticoid copepods, a study in the South Atlantic and Southern Ocean recorded 19 species of which 11 were restricted to particular regions, and eight widespread between ocean basins (Gheerardyn & Veit-Köhler 2009).

In polychaetes, sampling and analysis projects associated with CeDAMar have revealed both cosmopolitanism and cryptic speciation. Several species of small infaunal deposit-feeding spionids from abyssal depths are apparently distributed globally, based on examination of gross and ultra-structural morphology using scanning electron microscopy (Mincks et al. 2009; A. Glover unpublished data). Conversely, specimens of *Aurospio dibranchiata* Maciolek, 1981 from two central Pacific abyssal plain sites appear to be cryptic species based on 18S rRNA sequences, a normally highly-conserved gene (Mincks et al. 2009). A study of the distribution of multiple species of Southern Ocean abyssal polychaetes has revealed similar trends in terms of broad distributions of several species, based on morphology. Out of 70 Southern Ocean species studied in detail, 17 were shown to be cosmopolitan and only 13 apparently locally restricted to particular Southern Ocean sites (Schüller & Ebbe 2007). The remainder were at the very least broadly distributed, some between ocean basins (for example the Southern Ocean and North Atlantic).

A review of the distribution of protobranch bivalves in the east and west North Atlantic has revealed broadly distributed species at multiple bathymetric levels (McClain et al. 2009). Forty-three percent of the species studied were shared between the two ocean basins, of which 88% had overlapping depth ranges. The degree of apparent cosmopolitanism increased with depth, from 40% in bathyal regions to 60% in abyssal.

Systematic studies of deep-sea holothurians from the Galathea expedition revealed several cosmopolitan species in the abyss (Hansen 1975). Few taxa have been studied yet in detail using molecular methods, but the cosmopolitan species *Oneirophanta mutabilis* Théel, 1879 and *Psychropotes longicauda* Théel, 1882 have been recovered from multiple ocean basins. These species are characterised by large egg sizes up to 1 mm, which suggests lecithotrophic larvae or direct development (Ramirez-Llodra et al. 2005).

One of the more enigmatic abyssal groups is the Komokiacea, a group of soft-bodied formaminfera that produce large branching tests. A recent systematic review of komokiaceans from the Southern Ocean has revealed nine species, of which five are also present in the North Atlantic (Gooday et al. 2007).

Some foraminiferans apparently are truly cosmoplitan as they cannot be discriminated even with molecular genetic methods, indicating that gene flow is taking place from pole to pole (Lecroq et al. 2009). This global gene flow is difficult to imagine at first glance, and it may be confined to organisms with certain traits in their biology. Body size, which is inversely related to population size (that is, the smaller the organism is, the more individuals there are), plays an important role, and so do planktonic dispersal capabilities and the ability to survive long periods of famine. For example, the cosmopolitan species *Epistominella exigua* can live in substrata with organic carbon concentrations spanning orders of magnitude and episodic flux to small ephemeral patches on a seemingly homogeneous seabed (Lecroq et al. 2009). This flexibility is thought to facilitate gene flow even under marginal conditions.

In summary, available data are sparse yet support the view that both cosmopolitanism and basin endemism occur across a wide range of taxa in the abyss. These include species that exhibit direct development and bi-phasic life cycles where larvae can be carried by ocean currents. Evidence frommolecular genetics is now starting to challenge some of these apparent cosmopolitan distributions, but even if many abyssal species are cryptic, it is clear that gross morphologies, and in some cases fine ultrastructure, are highly conserved in the abyss.

This pattern may be a result of relatively rapid colonisation of the abyss from bathyal depths and subsequent slow rates of adaptive radiation, in response to relatively similar environmental conditions.

Studies of reproductive biology are extremely rare, and are required to find independent lines of evidence for species ranges. Polychaetes with clear direct-developing offspring have recently been recovered from an isolated, oligotrophic central Indian Ocean abyssal site that are apparently conspecific with specimens from both the north Atlantic and north Pacific. The simplification of a pattern where only species with larval stages are likely to be broadly dispersed is clearly being challenged, future studies involving physiological data and modelling of available habitats may yet provide the additional lines of evidence required to resolve the paradox of cosmopolitan abyssal species.

However, as so many animals in the abyss are rare, any distributional patterns have to be interpreted with great caution. "Endemic" species may just not have been found again in other locations, and all newly described species are by default "endemics". Conversely, many species considered to be cosmopolitan may have been misidentified, for example, through the use of identification keys not pertaining to the area. There is some indication that generally, distributional patterns as we interpret them from samples taken so far may represent extreme patchiness. The scale of this patchiness may be rather small (Kaiser et al. 2007), and we may have to change sampling strategies from large-scale coverage of entire ocean basins to concentrated sampling at a single site.

Evolution and Speciation in the Abyss

During the past several decades, much has been learned about patterns of species diversity in the deep sea and their potential ecological causes. However, we are only now beginning to explore the evolutionary processes that generated this rich and distinctive fauna. How and where did all these species originate? Currently, our entire understanding of evolution is based on patterns in other ecosystems.

Deep-sea mollusks were chosen for a study of deep-sea evolution because their basic taxonomy and biogeography is particularly well known. The ENAB project is testing models of evolution based originally on analyses of shell form within species arrayed along depth gradients (Etter & Rex 1990). This research suggested that most population differentiation occurred at intermediate depths in the narrow bathyal zone along continental margins, and that the abyss played only a minor role in promoting deep-sea biodiversity. However, it was not possible to determine whether bathymetric ranges in shell form represented evolved genetic differences or simply environmentally caused morphological differences.

New laboratory methods were developed to extract, amplify, and sequence mitochondrial DNA from specimens that had been fixed in formalin and then preserved in alcohol, sometimes for decades. The resulting patterns of genetic differentiation tended to confirm that the bathyal zone was an evolutionary hot spot (Etter et al. 2005). This research has now been expanded to examine very large-scale geographic variation in mollusks among deep-sea basins in the North and South Atlantic (Zardus et al. 2006). A variety of patterns has emerged including differentiation at great depths.

In the summer of 2008, the first deep-sea sampling expedition devoted exclusively to studying evolutionary patterns in the deep sea was performed. The objective was to collect fresh material in order to sequence both nuclear and mitochondrial genes. A broad range of genes is essential to verifying geographic patterns of differentiation. Fresh material also enables us to develop better primers to more effectively sequence genes in the vast amount of archived preserved material. Being able to use multiple genes adds a new dimension to evolutionary studies in the deep sea. Except for foraminiferans, there is virtually no fossil record of dep-sea assemblages to assist us in unravelling long-term adaptive radiation and the global spread of higher taxa. Instead, phylogenetic evolution must be inferred from molecular genetic data. For the first time, we now have broadly distributed material that is amenable to phylogeographic analysis. This will allow us to answer very fundamental questions, adding an evolutionary-historical perspective to our understanding of life in the deep sea. One of te most puzzling discoveries of this research so far is an apparent genetic break within eurybathic species at about 3,300 m, indicating that there is limited gene flow around this depth. This phenomenon not only occurs in mollusks, but was also reported for a widely distributed amphipod (France & Kocher 1996).

Nodule Mining and MPAs in the Pacific Abyss

Manganese nodules, or polymetallic concretions of iron and manganese hydroxides, can be abundant at the abyssal seafloor beneath regions of low to moderate ocean primary productivity (Ghosh and Mukhopadhyay 2000). In some regions, nodules may cover more than 50% of the seafloor and are potential mineral sources of copper, nickel, and cobalt. Manganese nodule mining is expected to occur in the abyss by the year 2025 and could ultimately be the largest scale human activity to directly impact the deep-sea floor (C.R. Smith et al. 2008b).

Thirteen pioneer investor countries and consortia have conducted hundreds of prospecting cruises to investigate areas of high manganese nodule coverage in the Pacific and Indian Oceans, especially in the area between the Clarion and the Clipperton fracture zones, which covers roughly 6 million km^2 and may contain 340 million tonnes of nickel and 265 million tonnes of copper (Ghosh and Mukhopadhyay 2000; Morgan 2000). Eight contractors are now licensed by the International Seabed Authority (ISA) to explore nodule resources and to test mining techniques within individual claim areas, each covering 75,000 km^2. In addition to harbouring mineral resources, abyssal Pacific sediments in the CCZ may also be major reservoirs of biodiversity (Glover et al. 2002).

However, it has been extremely difficult to predict the threat of nodule mining to biodiversity (in particular, the likelihood of species extinctions) because of very limited knowledge of (1) the number of species residing within areas likely to be perturbed by single mining operations, and (2) the typical geographic ranges of species within the nodule provinces (Glover & Smith 2003). During the CeDAMar field projects KAPLAN and NODINAUT, we used state of the art molecular and morphological methods to begin to evaluate biodiversity and species ranges of three key faunal groups in the abyssal Pacific nodule province: polychaete worms, nematode worms, and foraminiferans. Together, these groups can constitute more than 50% of faunal abundance and species richness in abyssal sediments (Smith & Demopoulos 2003), and represent a broad range of ecological and life-history types.

CeDAMar results indicate high, unanticipated levels of species diversity for all three sediment-dwelling faunal components studied at our individual sites E, C, and W. Based on morphological analyses, the Foraminifera contain at least 252 species at site E and at least 180 species at site C (Nozawa et al. 2006). Many of these species are new to science and appear not to have been collected elsewhere (Nozawa et al. 2006; C.R. Smith et al. 2008c). Based on DNA sequencing studies, the nematode worms also exhibit very high within-site diversity, with 73 molecular operational taxonomic units (or putative species) from only 97 sequenced individuals (C.R. Smith et al. 2008c). Because of a high ratio of one new species for every 1.3 individuals sequenced, the total nematode species richness is still grossly undersampled; we can be certain that far more species remain to be collected at each of our abyssal Pacific sites.

The polychaetes also exhibit very high within-site diversity for the families studied in detail; for example, Site E contains at least 48 polychaete species within 16 polychaete families (C.R. Smith et al. 2008c). A high abundance of apparently cryptic species found with our molecular studies indicates that earlier estimates of polychaete species richness within abyssal Pacific sites based on morphological studies, for example the 170 species from 3 m^2 by Glover et al. (2002), are likely to be low by at least a factor of two. We speculate that, even based on the relatively limited number of samples we have been able to analyse thus far, the total species richness of sediment-dwelling foraminiferans, nematodes, and polychaetes (a subset of the total fauna) at a single site in the CCZ could easily exceed 1,000 species (C.R. Smith et al. 2008c).

Our combined results for the foraminferans, nematodes, and polychaetes suggest that there is a characteristic fauna of the Pacific abyss, indicating that the abyss is not merely a sink of non-reproducing individuals transported from the continental margins (Rex et al. 200; C.R. Smith et al. 2008a). Many of the hundreds of species of Foraminifera identified from our samples appear to be restricted to, or at least characteristic of, the abyss (Nozawa et al. 2006; C.R. Smith et al. 2008c). Seventy of the 73 molecular operational taxonomic units (MOTUs) of nematodes appear to be new genera distinct from shallow-water genera, and thus may well have evolved in the abyss (C.R. Smith et al. 2008c). The molecular data for the polychaetes also indicate numerous cryptic new species in our KAPLAN abyssal samples, again suggesting that the abyssal polychaete fauna contains higher species diversity than previously appreciated, and may include numerous species evolved in the abyss. All of these results suggest that the central Pacific abyss harbours a specially adapted, diverse fauna distinct from the fauna of the continental margins. It seems very unlikely that all, or even many, species found in the CCZ abyss are protected from extinction by populations residing many thousands of kilometres away at much shallower depths on the continental margins (C.R. Smith et al. 2008a).

Although the data are still limited, there is significant evidence that community structure of the Foraminifera and polychaetes differ substantially on scales of 1,000–3,000 km across the CCZ. These apparent patterns of faunal turnover seem likely to be driven in part by the east to west decline in primary productivity thus the flux of food to the seafloor across the CCZ , but may also be driven in part by varying habitat heterogeneity (C.R. Smith et al. 2008c).

Using results from the KAPLAN and NODINAUT projects, CeDAMar helped to convene a workshop of experts to draft recommendations to ISA for the design of MPAs in the CCZ to conserve marine biodiversity and ecosystem structure and function in the region in the face of nodule mining. Based on sound scientific principles, it was recommended that a network of nine 400 km × 400 km protected areas (or "areas of particular environmental interest") be set up within the CCZ where mining would be prohibited (International Seabed Authority 2008, 2009). This network of protected areas would be stratified by regional variations in primary productivity and protect a total area of 1,440,000 km ^{2}, placing roughly 25% of the total CCZ management area under protection (International Seabed Authority 2008). The ISA is currently considering these recommendations. If implemented, these CeDAMar recommendations would initiate scientifically based conservation management in international waters, would establish the ISA as a leader in the application of modern conservation management, and would set a precedent for protecting seabed biodiversity, a common heritage of mankind, before the initiation of exploitive activities (International Seabed Authority 2008).

Natural History and Environmental Factors

Although we learned much about the faunal elements of abyssal benthos communities, we still know almost nothing about the natural history of abyssal animals or environmental factors structuring abyssal communities. To the human eye an abyssal plain looks uniform over hundreds of kilometres. Nonetheless, benthic communities are not nearly as homogeneous as originally thought. To abyssal animals, the habitat bears enough heterogeneity to cause species turnover even within a single ocean basin. However, we are just beginning to understand the scale of species turnover in abyssal plains.

In the deep Southern Ocean, the ANDEEP project has revealed patterns of biodiversity within different faunal groups, but we still do not know anything about the processes behind these biodiversity patterns. The ANDEEP follow-up International Polar Year project SYSTCO (system coupling) therefore focuses on coupling processes between atmosphere, water column, and deep-sea floor near the Polar Front and in the abyssal Weddell Sea and includes ecological questions and investigations of the role of deep-sea fauna in trophodynamic coupling and nutrient cycling in oceanic ecosystems.

Speciation in the Abyss

On an evolutionary scale, the same gap in our knowledge becomes apparent. We know very little about speciation in the abyss, and we are just now beginning to gain insights into the origin of the abyssal fauna and the very high diversity of abyssal benthic communities. Especially for soft-bodied organisms that leave no fossil record, molecular clocks have to be developed to reconstruct their evolutionary history. ENAB has developed novel techniques which are promising for future research.

Abyssal Species Numbers and Taxonomy

We will probably never know the true number of species in the abyss. The research area is far too large to be sampled adequately considering how heterogeneous this habitat turns out to be and how high the percentage is of rare species which have been recorded from just one site, often also by just one individual among thousands. Nevertheless, with knowledge gained continuously, scientists continue to try to reach better and better estimates.

The remarkable gain of knowledge about the abyssal benthos, notwithstanding the taxonomic impediment which brought about the birth of CeDAMar, is still apparent. We are still facing an overwhelming amount of species awaiting formal description and a scarcity of specialists to do the task. Taxonomic intercalibration, which has come a long way during CeDAMar, will have to continue as we have just scratched the surface. Molecular genetic and morphological methods will have to be integrated in a continuing effort to understand each other and communicate.

Moving On

Although public awareness about the deep sea has risen a great deal during CeDAMar, the abyss is still perceived by most people as a somewhat remote part of the planet, not affecting humankind in any way worth mentioning, and the research is still felt to be somewhat academic.

However, the abyss is on its way to become a resource for human exploitation very quickly. Industrial harvesting of manganese nodules may become a reality before most of us notice. Necessary technology is far advanced, largely unnoticed by anybody other than those directly involved. Even before man-made gear enters this still pristine environment, it is quite possible that the abyssal seafloor, which accounts for the largest area on the planet, may warrant our close attention because biogeochemical cycles of the seafloor have a strong influence on the global climate and climate change.

Climate warming is expected to increase regional sea surface temperatures and thermal stratification in low to mid-latitudes, yielding reductions in nutrient upwelling (C.R. Smith et al. 2008b; K.L. Smith et al. 2009). These changes will in turn alter the quantity and quality of food flux from the euphotic zone to the abyssal seafloor. CeDAMar studies suggest that resulting long-term declines in POC flux to the abyss will cause reductions in the abundance and biomass of benthic fauna, and yield reductions in species diversity and body size over large regions, such as in the equatorial Pacific. Substantial shifts in the taxonomic composition of abyssal assemblages, especially the megafauna, are also expected, as well as changes in basic ecosystem functions at the seafloor, such as organic carbon burial and calcium carbonate mineralisation.

Climate induced reductions in abyssal food flux over large areas, such as the equatorial Pacific biodiversity hot spot, have the potential to cause regional species extinctions as populations are reduced below reproductively viable levels (Rex et al. 2005; C.R. Smith et al. 2008b). Because abyssal ecosystems are so sensitive to the quantity and quality of sinking food material from the upper ocean (C.R. Smith et al. 2008b; K.L. Smith et al. 2009), impacts on the abyss must be considered in predicting the effects of climate warming and eco-engineering (for example ocean fertilization to mitigate climate change) on the biodiversity and ecological functioning of ocean ecosystems.

What Needs to Be Done?

When the first Census has ended, keeping the momentum of global collaboration has to become our first action item. One idea might be to establish an international consortium supported by national funding agencies to identify important questions that most urgently need answers. Funding for taxonomists and molecular biologists needs to be secured in the long term to truly overcome the taxonomic impediment. Sampling strategies need the same global perspective as the Census to avoid falling back into competition among nations or institutions for the most attractive results.

Innovative methods will have to be adopted for the exploration of life in the abyss, for example, in situ experiments that might tell us something about the biology of abyssal organisms, and autonomous vehicles that can travel along abyssal plains to collect data over large distances and areas. The technically challenging development of suitable instruments and research with such methods will require substantial additional funding which will be granted only if the general public gets involved and educated. Societal acceptance of deep-sea research is still measured by that of astronomy. Allocating public funds to investigate other planets, stars, and even galaxies, immeasurably farther away from human reach, is questioned by few, in contrast to investigating the portion of surface of our own planet which happens to be covered with water.

Exhibitions and trade fairs related to boating and diving lately included small individual submarines for pleasure, designed to dive to about 100 m, driven by the owners themselves. Although these submarines are targeted for a very wealthy clientele, they may perhaps raise awareness for the benthic environment in a different and more direct way than anything we can offer through the media.

Outlook and Conclusions

The return to a more holistic perspective is perhaps a logical process following nearly a century of specialisation and focus on smaller and smaller details of an ecosystem which, as we gathered more and more facts, seemed to become more and more difficult to comprehend. We may have reached a time that is right for taking a step back and looking at whole systems from different viewpoints, realising how they all overlap and complement each other. If one could understand which factors regulate the presence of species in a given area and which factors regulate the absolute and relative abundance of these species, then one would understand much of the functioning of the ecosystem as a whole. The evaluation of biodiversity – defined as the variety and variability of genomes, populations, species, communities, and ecosystems in space and time (Heywood 1995) – continues to be a central theme in biology and conservation.

When the scientific scope of CeDAMar was planned, exclusion of continental margins, seamounts, and chemosynthetic environments was deliberate. Only through focusing on a few of the major abyssal basins of the global ocean was it possible to achieve any tangible results in the limited timeframe of the Census. Exploring the relations of the ecosystem "abyssal benthos" with neighbouring systems is a logical second step to be undertaken in the future.

Several habitats possibly interacting with the abyssal benthos come to mind, most obviously the continental margins; on an even larger scale, an integration of water column and benthos research is a desirable goal. To be able to gain more complete insights both spatially and temporally, the abyss must be integrated into ocean observing systems.

Although the rate of discovery of new species is intimidating, it is not equally large for all organisms. Specialists do not expect much beyond 10% new species, for example, of mollusks (whereas for others such as nematodes the rate may be about 90%). Several organisms have been found to be widespread, for example, on either side of the Atlantic Ocean or in both Polar seas. Although genetic investigations have to confirm these patterns based on morphology, we may eventually come to a realistic estimate of the number of species in the abyss. "Singletons", those species known from only one specimen, may eventually be recaptured at the original site or even elsewhere, and the recapture rate may be a good proxy for species richness.

Within the past 150 years, we have learned to look at the abyss through different lenses. The unfathomed depths turned from a mythical place inhabited, at best, by fearsome creatures waiting to attack the unwary seaman, to an integral part of our planet filled with a dazzling variety of life, well adapted to its environment and of unsuspected beauty and grace.

There are still many more questions than answers. CeDAMar research has lifted some of the mysteries, and the facts are even more fascinating than the myths, to scientists as well as the general public. There is much hope among deep-sea scientists that CeDAMar, together with other deep-sea projects within the Census, acted as a spark for ongoing research in the decades to follow.

Technology: Pushing Backwards the Limits to Knowledge

Despite the best possible combinations of sensors and technologies, a limit to progress will remain related to accessing the appropriate sections of the global ocean using suitable research ships. As we develop AUV-based approaches we can improve the rates of exploration and discovery. Because AUVs are decoupled from ships' positions and motions (unlike ROVs) and do not put human lives in harm's way at the bottom of the ocean (unlike manned submersibles), we will be able to expand our systematic exploration surveys to higher and higher latitudes. Furthermore, a new generation of AUVs is now under construction that will have ranges in excess of 1,000 km and could ultimately span entire ocean basins. At this level, and with artificially intelligent control systems, we can imagine a day when long-range AUVs will be able to conduct first pass investigations along entire sections of the global ridge crest, interrogating, processing, and interpreting their data underway, to allow second pass mapping and, potentially, even photographic surveys of the seafloor. Although such approaches may seem like so much science fiction, preliminary algorithms to interpret hydrothermal field data on the fly have already been trialled (for example on the MAR) and a first long-range mid-ocean ridge ship-free AUV cruise has already reached the planning stage

Human Footprints in Deep-Water Chemosynthetic Ecosystems

Deep-sea ecosystems support one of the highest biodiversities of the Planet as well as important natural and mineral resources. In the last decades, the depletion of resources in the upper layers of the oceans together with technological development has fuelled the increasing interest of industry to explore and exploit deep-sea resources. Industries such as mining, hydrocarbon extraction, and fishing are working at great depths (below 1,000 m), and some of these activities affect chemosynthetic ecosystems. We briefly discuss below only anthropogenic impacts that have been addressed by ChEss scientists (i.e. mining, hydrocarbon extraction, and trawling), but other impacts such as litter accumulation or climate change are also important.

Major Known Impacts

Probably the most important industry potentially affecting chemosynthetically driven habitats and their fauna is mining for precious metals on seafloor massive sulfide deposits (SMS) from vents. SMS contain significant quantities of commercially valuable metals, such as gold, silver, copper, and zinc (Baker & German 2009). Although at a very early stage of exploration, the SMS industry is already extremely active, with two major companies (Nautilus Minerals and Neptune Minerals) developing plans for exploitation in national waters of Papua New Guinea and New Zealand. ChEss scientists have participated in the environmental impact assessment conducted by Nautilus Minerals, who have produced an environmental impact statement for mining activities in Papua New Guinea, and considerable baseline research has been done (Levin et al. 2009b; Erickson et al. 2009; see EIS at Although it seems highly likely that economically viable extraction of sulfides from the deep-sea floor may begin within the next decade, the true nature of mining impact on the ecosystem is still mostly unknown.

There are several potential environmental effects of mining that are of concern to some stakeholders. These include direct physical damage to the seabed at the operation site and the surrounding area; potential extinction of isolated populations; production of sediment plumes and deposition of sediment, which will affect marine life by smothering or inhibiting filter feeders; alteration of fluid-flow paths at a vent, on which the benthic, often sessile, vent fauna rely; noise pollution; wastewater disposal; and equipment failure which may result in leakage (ISA 2004; Van Dover 2007). ChEss has been instrumental in instigating key collaborations including all stakeholders, conducting baseline research, participating in workshops and discussion fora, and contributing to the "Code for Environmental Management of Marine Mining" (IMMS 2009) and the International Seabed Authority's guidelines for environmental baselines and monitoring programs (International Seabed Authority 2004; Van Dover 2007), which both serve to monitor and mitigate the potential effects of deep-sea mining.

Another important extractive industry is oil and gas exploitation. Seep communities often coincide with large subsurface hydrocarbon reservoirs and the outcropping of gas hydrates. Thus they may also be susceptible to damage from oil and methane exploration and extraction activities in the future. One of the most intensive areas of deep-water oil industry activity is in the Gulf of Mexico, leased and monitored by the US Minerals Management Service. Collaborations between scientists and the Minerals Management Service in the Gulf of Mexico are well established (Fisher et al. 2007; Roberts et al. 2007), and exploration for new sources of hydrocarbons often coincides with the discovery of new biological communities. As new high-density biological communities are discovered, the Minerals Management

Service establishes "mitigation areas" to prevent oil industry activity from impacting these sensitive areas. ChEss has also established strong collaborations with Fondation Total to promote science in chemosynthetic ecosystems.

Finally, ChEss scientists have been involved in issues related to deep-water trawling that affect chemosynthetic sites. Recent exploration of the eastern margin of New Zealand provided not only the first descriptions of seep communities in the region but evidence that seep sites have been subjected to bottom fishing (Baco-Taylor et al. 2009), with visible accumulations of coral or vesicomyid shell debris, lost trawl gear, or trawl marks in seep adjacent sediments. Similarly, signs of trawling have been found on the California margin at 500 m depth (Levin et al. 2006) and on the Chilean margin, where the commercially fished Patagonian toothfish is closely associated with seeps (Sellanes et al. 2008).

Marine Protected Areas

As vent sites have become the focus of intensive, long-term investigation, both governmental and non-governmental organisations have been discussing the need to introduce appropriate measures that combine preservation of habitat with scientific interests, tourism, and, potentially, mining (Mullineaux et al. 1998; Dando & Juniper 2000; Santos et al. 2003).

The ChEss project has been an active participant in the planning of biogeographically representative networks of sites of interest for conservation and continued scientific research. ChEss scientists have contributed also to the Convention of Biological Diversity effort, to produce scientific criteria and guidance for the identification of ecologically or biologically significant marine areas and the designation of representative networks of marine protected areas (MPAs) (Convention on Biological Diversity 2009). So far, ecological reserves and/or MPAs have been proposed at the Mid-Atlantic Ridge and the East Pacific Rise, and have already been established on the Juan de Fuca Ridge.

At the MAR, three sites – Lucky Strike, Menez Gwen, and Rainbow – were proposed to be included on the network of MPAs of The Convention for the Protection of the Marine Environment of the North-East Atlantic (OSPAR) maritime region V (the wider Atlantic). In the New Zealand region, some hydrothermal vent sites are currently protected from bottom trawling by a Benthic Protected Area. Although licenses that have been issued for exploratory mineral mining overlap with this area, it is expected that collaborations among stakeholders will instigate new prospects for conservation.

The scientific community has recognised also the potential impact of continuous research activities at certain sites. This led to the "Statement of Responsible Research Practices at Hydrothermal Vents", developed by InterRidge with the collaboration of ChEss scientists (Devey et al. 2007). Following on those steps, the OSPAR convention also proceeded with the design of a Code of Conduct for Research on Hydrothermal Vents, which included the usefulness of MPAs (OSPAR draft background report 2009).

Conclusions

Under the ChEss project, the scientific community has expanded our knowledge beyond the vent biogeographic regions recognised in the early twenty-first century (Van Dover et al. 2002). New technologies have been developed, making available the necessary tools to explore and investigate remote habitats of difficult access in a very efficient way, resulting in the discovery of new sites in all oceans on Earth.

The number of species described from vents, seeps, whale falls, and OMZs is constantly growing, providing essential data to understand global biodiversity and phylogenetic links among habitats to refine existing biogeographic provinces and define new ones.

Indeed, Bachraty et al. (2009) have recently addressed the biogeographic relationships among deep-sea vent faunas at a global scale using statistical analysis of comprehensive vent data. They delineate six major hydrothermal provinces and identify possible dispersal pathways.

Furthermore, detailed ecological investigations (for example trophic structure, reproduction, larval ecology) of known sites have resulted in a better understanding of ecosystem functioning and the role played by the environment in shaping deep-water chemosynthetic communities.

As already given some of the key contributions the ChEss project participants have made toward a better global understanding of the distribution of chemosynthetic environments, the species that inhabit them, and some underlying ecological processes. Nevertheless, there are important geographic gaps in the global chemosynthetic biogeographic puzzle that remain to be explored and gaps to knowledge that remained unanswered in 2010. The momentum created by the Census and ChEss initiatives has promoted major international collaborations and strengthened existing ones.

This synergy between laboratories around the globe, sharing expertise and resources in joint research projects, will continue beyond 2010 as one of the main legacies of ChEss.

The scientific results obtained are crucial for the development of conservation and management options in an ecosystem that is already affected by human activities.

3

A Global Census of Marine Microbes

The oceans abound with single cells that are invisible to the unaided eye, encompassing all three domains of life – Bacteria, Archaea, and Eukarya – in a single drop of water or a gram of sediment. The microbial world accounted for all known forms of life for more than 80% of Earth's history. Today, microbes continue to dominate every corner of our biosphere, especially in the ocean where they might account for as much as 90% of the total biomass (Fuhrman et al. 1989; Whitman et al. 1998). Even the most seemingly inhospitable marine environments host a rich diversity of microbial life. For the past six years, microbial oceanographers from around the world have joined the effort of the International Census of Marine Microbes (ICoMM, Box 12.1) to explore this vast diversity. In this chapter we provide a brief history of what is known about marine microbial diversity, summarise our achievements in performing a global census of marine microbes, and reflect on the questions and priorities for the future of the marine microbial census.

From the time of their origins, single-cell organisms – initially anaerobic and later aerobic – have served as essential catalysts for all of the chemical reactions within biogeochemical cycles that shape planetary change and habitability. Marine microbes carry out half of the primary production on the planet (Field et al. 1998). Microbial carbon re-mineralisation, with and without oxygen, maintains the carbon cycle. Microbes account for more than 95% of the respiration in the oceans (Del Giorgio & Duarte 2002). They control global utilisation of nitrogen through N_2 fixation, nitrification, nitrate reduction, and denitrification, and drive the bulk of sulfur, iron, and manganese biogeochemical cycles (Kirchman 2008; Whitman et al. 1998). Marine microbes regulate the composition of the atmosphere, influence climate, recycle nutrients, and decompose pollutants. Without microbes, multicellular animals on Earth would not have evolved or persisted over the past 500 million years. Measuring microbial diversity in a broad range of marine ecosystems will facilitate quantification of the magnitude and dynamics of the microbial world and its stability through space and time. The phylogenetic and physiological diversities of microbes are considerably greater than those of animals and plants, and microbial interactions with other life-forms are correspondingly more complex (Pace 1997). Measuring marine microbial diversity and determining corresponding associated functions will thus provide a wealth of information about specific microbial processes of great significance such as wastewater treatment, industrial chemical production, pharmaceutical production, bioremediation, and global warming. Examining the relationships between microbial populations and whole communities within their dynamic environment will allow us to formulate better the definition of what constitutes an ecologically relevant species in the microbial world.

Molecular methods rely upon measures of genetic similarity to describe operational taxonomic units (OTUs). Statistical treatments can use the relative number of distinct OTUs to estimate diversity, but these inferences do not translate directly into numbers of microbial species. Microbiologists have not reached consensus on the definition of microbial species using either molecular or phenotypic approaches. However, ecological concepts of microbial species based upon molecular data will inform theoretical applications and guide solutions to major challenges facing science and human society.

Microbial Diversity and Abundance

The reliance upon traditional cultivation and staining techniques led to gross underestimates of microbial abundance and species richness in both oceanic and terrestrial environments (Jannasch & Jones 1959; Zimmermann & Meyer-Reil 1974; Hobbie et al. 1977). The application of fluorescence-based microscopy coupled with DNA staining methods revealed the great "plate count anomaly", which posits that microbiologists have underestimated microbial abundances by at least three orders of magnitude. Instead of a mere 100 cells per millilitre of seawater, nucleic-acid staining technology showed the number of bacteria in the open ocean exceeds 10^{29} cells, with average cell concentrations of 10^{6} per millilitre of seawater (Whitman et al. 1998). In marine surface sediments, cell abundances are 10^{8}–10^{9} per gram, and even in the greatest depths of the subsurface seabed, more than 10^{5} cells per gram are encountered (Jørgensen & Boetius 2007). The ocean also hosts the densest accumulations of microbes known on Earth, reaching 10^{12} cells per millilitre, like the photosynthetic mats thriving in hypersaline environments, and the methanotrophic mats of anoxic seas, resembling ancient microbial assemblages before the advent of eukaryote grazers (Knittel & Boetius 2009). Archaeal cell abundances rival those of bacteria in certain parts of the ocean and the seabed, and microbial eukaryotic (protistan) densities vary widely from tens of cells per litre to bloom conditions that can surpass 10^{6} cells per millilitre of seawater.

As of 2010, cultivation-based studies have described more than 10,000 bacterial and archaeal species nd an estimated 200,000 protistan species (Corliss 1984; Lee et al. 1985; Patterson 1999; Andersen et al. 2006). Cultivation-independent studies that rely upon molecular methods such as the sequencing of 16S ribosomal RNA (rRNA) genes show microbial diversity to be approximately 100 times greater (Pace 1997). With each new molecular survey, this window on the microbial world has increased in size.

Challenges of a Microbial Census

The ocean covers 70% of the Earth's surface (an estimated volume of about 2×10^{18} m^{3}) and has an average depth of 3,800 m. Strategies for conducting a census must consider the enormous geographical area to be surveyed, an almost unimaginable number of cells, and the impact of spatial gradients and temporal shifts on microbial assemblages. In fact, before ICoMM, little was known about global patterns in microbial communities. Basic questions such as "is there a distinct difference between pelagic and benthic microbial communities?" or "what is the temporal turnover in microbial cells between two sampling dates?" profoundly influenced our sampling strategies.

Contemporary molecular approaches typically use rRNA sequences as proxies for the occurrence of different microbial genomes in an environmental DNA sample (coding regions for functional genes can also provide information about microbial population structures). However, the expense of conventional DNA sequencing has constrained the number of homologous sequences that microbial ecologists typically collect to describe community composition. Relative to the number of microbes in most samples, these surveys superficially describe microbial community structures. There are more than 10^{8} microorganisms in a litre of seawater or a gram of soil (Whitman et al. 1998). Few studies collect more than 10^{4} sequences, which correspond to 0.01% of the cells in a litre of seawater or a gram of soil. The detection of organisms that correspond to the most abundant OTUs or species equivalents requires minimal molecular sampling, whereas the recovery of sequences from rare taxa that constitute the "long tail" of low abundance organisms in taxon rank–abundance curves demands surveys that are orders of magnitude larger.

As an alternative to analysing nearly full-length polymerase chain reaction (PCR) amplicons of rRNA genes from environmental DNA samples, short sequence tags from hypervariable regions in rRNAs (pyrotags) can provide measures of diversity (species or OTU richness) and relative abundance (evenness) of OTUs in microbial communities. When combined with the massively parallel capacity of "next generation" DNA sequencing technology that allows for the simultaneous sequencing of hundreds of thousands of templates (Margulies et al. 2005), it becomes possible to increase the number of sampled gene sequences in an environmental survey by orders of magnitude (Huber et al. 2007; Sogin et al. 2006). Enumerating the number of different rRNA pyrotags provides a first-order descrition of the relative occurrence of specific microbes in a population. The highly variable nature of the tag sequences and paucity of positions do not allow direct inference of phylogenetic frameworks.

However, when tag sequences are queried against a comprehensive reference database of hypervariable regions within the context of full-length sequences, it is possible to extract information about taxonomic identity and microbial diversity. Initial tests of this innovative technology examined the microbial population structures of samples from the meso and bathypelagic realm of the North Atlantic Deep Water Flow and two diffuse flow samples from Axial Seamount on the Juan de Fuca ridge off the west coast of te United States (Sogin et al. 2006). These initial data sets led ICoMM investigators to the discovery of the "rare biosphere", a rich diversity of novel, low-abundance populations and dormant or slow growing microbes.

A singe litre of seawater, on average containing 10^{8}–10^{9} bacteria, represents about 20,000 "species" of bacteria, a number that is one or two orders of magnitude higher than estimated earlier (Venter et al. 2004). When plotted on a two dimensional x–y microbial rank distribution diagam, this species-richness shows an extraordinarily long tail, the long tail including low-abundance taxa, many of which represent types of microbes that have never been seen before. Huber et al. (2007) extended this approach to the Archaea, also targeting the V6 16S rRNA hypervariable region and reported species richness estimates to be on the order of 3,000 "species" per litre of seawater. Amaral-Zettler et al. (2009) developed a tag sequencing strategy for the V9 hypervariable region of the 18S rRNA gene in eukaryotes and determined that estimates of microbial eukaryotic (protist) species richness can be on the order of magnitude seen in the archaeal domain but may be an order of magnitude lower in more extreme environments such as Antarctic waters.

The International Census of Marine Microbes subsequently adopted this pyrotag strategy in a coordinated microbial census of samples from globally distributed marine environments. A study of lipid molecular structures from marine microbes complements the pyrotag survey. The database MICROBIS serves information to ICoMM, and its website provides access to this information including the capacity to retrieve contextual data information for all samples

The database VAMPS (Visualisation Analysis of Microbial Population Structures, and its links to MICROBIS provide full access to the pyrotag sequences, the contextual data, analytical and graphical tools for comparing microbial population structures for different sites, search tools for locating sequences in each of our samples, descriptions of community composition at taxonomic ranks of phyla, class, order, family, or, when possible, genus for all samples, and rarefaction and diversity analyses for all of ICoMM's data. The geospatial breadth of pyrotag and lipid data for this global study of microbes in the world's oceans has been depicted. It includes a subset of more than 18 million DNA sequence reads distributed among 583 bacterial, 120 archaeal, and 59 eukaryotic datasets from a larger dataset of >25 million sequences from >1,200 samples.

The samples represent all major oceanic systems including the Atlantic, Pacific, Arctic, Southern, and Indian Oceans, and sediment and water samples from estuaries to deep-water environments including vents and seeps, seamounts, corals, sponges, microbial mats and biofilms, and polar regimes. It has been described the origin of samples, targeted domains, project descriptions, and relevance to other Census ocean-realm projects. Here we present a broad-brush synthesis of our data emphasising the most abundant pyrotags recovered from our surveys. Although a comprehensive synthesis of these data lies beyond the scope of this chapter and the highlights that follow offer a glimpse into novel insights that will soon emerge from this international study of microbial community structures of the world's oceans. More detailed meta-analyses will frame the bulk of ICoMM's working groups during 2010.

The "Abundant Biosphere"

The pie charts summarise the most abundant tags in our bacterial, archaeal, and microeukaryotic datasets respectively. As expected from the work of S.J. Giovannoni in the Sargasso Sea (Giovannoni et al. 1990), pyrotags corresponding to á-Proteobacteria and specifically SAR11 represented the most abundant organisms (primarily in planktonic samples) in our global survey. This heterotrophic á-Proteobacterial lineage plays a critical role in the cycling of carbon, nitrogen, and sulfur and accounts for approxmately 25% of the biomass and 50% of the cell abundance in the ocean. More recently, researchers discovered that members of this group of bacteria contain proteorhodopsin, which potentially enables the harvesting of energy from light (Fuhrman et al. 2008; Giovannoni et al. 2005).

The presence of this clade in different habitats including coastal waters, seamounts, polar waters, and the open ocean (not shown) reflects its ubiquity in the marine pelagic environment. The 20 most abundant tags in our bacterial analyses also include members of the Rhodobacteraceae. One member of this group, *Roseobacter* sp., is cultivable by adding extracts of algal secreted organic matter to the medium (Mayali et al. 2008). The worldwide association of *Roseobacter* with algal blooms suggests it has a role in controlling bloom outbreaks.

The most abundant tag sequence derived from a photosynthetic bacterium belonged to a member of the Prochlorales (Cyanobacteria) and shares 100% V6 rRNA region sequence identity with the cultivar *Prochlorococcus marinus*. The picocyanobacteria (smaller than 2 μm) *Prochlorococcus* spp. along with *Synechococcus* spp. dominate the oceans with cell numbers of up to 10^5–10^6 per millilitre (Heywood et al. 2006; Scanlan et al. 2009). Collectively they contribute up to 50% of oceanic primary production (Li 1994). Cyanobacteria represent an ancient group of organisms. These inventors of oxygenic photosynthesis drove the oxygenation of the Earth's atmosphere 2.5 billion years ago. The evolution of aerobic Bacteria and Archaea made possible the origins of plants and animals about 0.5 billion years ago when the oxygen concentration in the atmosphere reached its present-day level. Today, Cyanobacteria produce about 50% of the oxygen on Earth. Most Cyanobacteria occur in marine communities (Garcia-Pichel et al. 2003).

Members of the phylum Crenarchaeota dominated the Archaeal pyrotag surveys. Microbial ecologists originally thought that all Crenarchaea represented extremophiles, until the discovery of their ubiquity in everyday marine and terrestrial environments (DeLong 1992; Fuhrman et al. 1992; Simon et al. 2000). Crenarchaeotal abundances can exceed bacterial abundances below 100 m depth in the ocean where they are metabolically active and can contribute to the oceanic carbon cycle (Herndl et al. 2005). In the Arctic, Marine Group III Euryarchaeota can dominate the deep water masses such as the deep Atlantic Layer in the central Arctic Ocean (Galand et al. 2009b). Sequences related to this group represented the second most commonly encountered pyrotags in our global archaeal dataset. In many cases, we only detected other abundant archaeal pyrotags in specific samples. For example, the methanogens Methanosarcinales occurred primarily in Lost City Hydrothermal Vents, whereas *Archaeglobus*- and *Methanococcus*-related tags specifically associated with sulfide chimneys.

The pyrotag studies showed that dinoflagellates dominated most eukaryotic microbial communities. Members of the dinoflagellates include phototropic and heterotrophic representatives and many co-occur with or may be responsible for harmful algal blooms, making them commercially and ecologically important. The high frequency of dinoflagellate tags likely reflects bias introduced by the very large copy number of rRNA genes in the genomes of most dinoflagellates (as many as 12,000 copies in species such as (Zhu et al. 2005)). Indeed, many of the tags recovered among the top 20 most abundant microbial eukaryotes included members of the picoeukaryotes (0.2–2 μm in size) that numerically dominate, but tend to have lower copy numbers of their 18S rRNA genes. The diversity of these Lilliputians of the protist world was only first recognised at the beginning of the twenty-first century (Díez et al. 2001; López-García et al. 2001; Moon-van de Staay et al. 2001).

The top most abundant tag among our eukaryotic datasets displayed 100% identity with the sequence of an unclassified dinoflagellate within the group that Gast et al. (2006) first identified in the Ross Sea, Antarctica. In some cases these cells occur at densities up to 29,000 cells per litre. Equally intriguing, our global analyses of eukaryotic tags revealed this pyrotag also occurs in the Arctic, Pacific, and Atlantic Oceans (from the Caribbean to the Gulf of Maine), the Framvaren Fjord in Norway and the Black Sea. Whether this tag represents the same cosmopolitan species or closely related ecotypes that extend over the globe remains unknown.

Because of differences in their gene copy number in different taxa, the relative abundance of eukaryotic pyrotags does not reflect the number of cells in a sampled environment. However, these data provide important taxonomic information at the species level for many morphologically rich eukaryotic microbes including dinoflagellates (for example and for members of the picoeukaryotes such as that are harder to distinguish morp The "Rare Biosphere"

In the microscopic realm, ICoMM's sampling of many diverse marine ecosystems has reinforced the concept of a ubiquitous rare biosphere (Pedrós-Alió 2006; Sogin et al. 2006). This forces us to reconsider the potential feedback mechanisms between shifts in extremely complex microbial communities and global change, as well as how microbial communities and the genomes of their constituents change over evolutionary timescales. Minor population members may serve as functional keystone species in microbial consortia, or they might be the products of historical ecological change with the potential to become dominant in response to shifts in environmental conditions, for example when local or global change favours their growth. The absence of information about the global distribution of members of the rare biosphere makes it impossible to ascertain if they represent specific biogeographical distributions of bacterial taxa, functional selection by particular environments, or cosmopolitan distribution of all microbial taxa the "everything is everywhere" hypothesis.

Data from recent pyrosequencing efforts, however, are beginning to shed light on this topic. For example, compared pyrotags from Arctic Ocean samples. These samples included both surface and deep waters, as well as winter and summer samples from different locations. When they clustered the samples, they observed nearly identical patterns for all abundant sequences (more than 1% of all tags), or only rare sequences (less than 0.01% of all tags). This indicates that in this system, the rare OTUs have the same biogeography as the abundant OTUs. This opens up two possibilities: either the pyrosequencing approach is only targeting the most abundant of the rare biosphere and the actual tails are even longer; or the rare OTUs must be a dynamic lot, able to grow and experience losses due to predation and viral attack in some intriguing and unknown way.

To explore whether high-abundance taxa represent physiologically active populations whereas low-abundance pyrotags represent less active or dormant microbes, Hamasaki et al. (2007) applied the bromodeoxyuridine (BrdU, thymidine analogues for detecting *de novo* DNA synthesis) magnetic bead immunocapture method to examine both the abundant and rare members of the microbial community. They applied this technique to surface seawater samples from four stations along a north–south transect in the South Pacific taken from November 2004 to March 2005 during the KH-04-5 cruise of the R/V *Hakuho-maru* (Ocean Research Institute, the University of Tokyo and JAMSTEC). They incubated subsamples on board after adding BrdU and then compared the pyrotag microbial community structures in treated and untreated samples. Their results indicated that some high-abundance taxa incorporated BrdU whereas others did not, and some rare taxa represented only by singletons in the resulting tag dataset also took up BrdU. More importantly, some BrdU-labelled taxa were detected only in the BrdU-labelled fraction but were absent in the untreated fractions. These results suggested that the rare biosphere is not restricted to physiologically inactive populations but can include taxa involved in biogeochemical cycles. This study further illustrates that there may be dynamic exchange between abundant and rare microbial populations.

High Archaeal Microdiversity in the Lost City Hydrothermal Field

The Lost City Hydrothermal Field on theMid-Atlantic Ridge is the first deep-sea environment discovered where exothermic water-rock reactions in the sub-seafloor and not magmatic sources of heat drive hydrothermal fluid flow. These reactions create a combination of extreme conditions never before seen in the marine environment: the venting of high-pH (from 9 to 11), warm 40–91 °C) hydrothermal fluids with high concentrations of hydrogen, methane, and other hydrocarbons of low molecular mass. The Lost City Hydrothermal Field may thus represent a new type of life-supporting system in the deep sea. Mixing of the warm, high-pH fluids with seawater precipitates carbonate that drives the growth of chimneys that tower up to 60 m above the seafloor. These carbonate towers, some of which remain active for thousands of years, house extensive microbial biofilms that are dominated by a single group of Archaea, the *Methanosarcinales*. However, multiple lines of evidence including morphology, and evidence or diversity of specific genes such as the genes involved in N_2 fixation and methane production and anaerobic oxidation, indicate physiological diversity within this *Methanosarcinales*-dominated biofilm.

Brazelton and colleagues correlated archaeal and bacterial V6 tag distributions with isotopic ages of carbonate chimneys collected from the Lost City Hydrothermal Field spanning a 1,200 year period. Clear shifts in the archaeal and bacterial communities were evident over time, and many of the shifts featured rare sequences that dramatically increased in relative abundance to become dominant in older chimneys.

These results indicate that some organisms can remain rare at a location for many years before "blooming" and becoming dominant when the environmental conditions allow. Furthermore, the very low overall diversity of the Lost City chimneys revealed that each of the dominant archaeal and bacterial sequences represented one member of a large pool of similar but much rarer sequences. For example, the most abundant archaeal sequence was more than 90% similar to 1,771 different sequences clustering into 517 operational taxonomic units at 97% sequence similarity, all of which were too rare to be detected by clone library sequencing.

Further work in the Baross laboratory has shown that this microdiversity in the V6 region is correlated with microdiversity in a more variable marker, the intergenic transcribed spacer (ITS) region, indicating that it is not generated by pyrosequencing error and that the archaeal population contains even more microdiversity than represented by the 1,771 variants detected in the V6 region. These results confirm that there are many rare species of *Methanosarcinales* that had not been previously identified from 16S rRNA gene clone libraries that likely represent multiple "ecotypes" within the biofilm and that the ecotype composition changes depending on the age of the carbonate structure.

Rapid Temporal Turnover in Sands of the North Sea Island of Sylt

Permeable sandy sediments play a critical role in the recycling of carbon and nitrogen, and act as natural filters that may concentrate microorganisms, nutrients, and organic matter on the extensive continental shelves. Despite the importance of such ecosystems, the extent of microbial diversity and how microbes respond to environmental, spatial, and temporal changes (such as global warming, ocean acidification, and various anthropogenic effects) are still mostly unknown.

Using a 454 massively parallel pyrotag sequencing strategy to describe microbial diversity in temperate sandy sediments from the North Sea island of Sylt, Ramette and colleagues obtained between 5,000 to 19,000 unique types of bacteria in each gram of sand (A. Gobet, S.I. Böer, J.E.E. van Beusekom, A. Boetius and A. Ramette, unpublished observations). Rarefaction analyses suggest that the OTU richness of sand-associated bacterial communities significantly exceeds the diversity of water column communities from the same environment. The OTU richness also changed dramatically over a few centimetres of sediment depth or between any two consecutive sampling times, with up to 70–80% community turnover. Those remarkable, non-random shifts in community composition may reflect responses to variation of many environmental/biogeochemical parameters (for example temperature, nutrients, pigments, production of extracellular enzymes) at the study site. The reservoir of highly diverse low abundant bacterial types might include taxa that become abundant in response to environmental differences in this system.

The comparison of diversity patterns at different taxonomic levels indicated that community shifts occurred at broad taxonomic levels, but that fine-scale patterns in community composition were mostly responsible for the large community turnover observed over sediment depth and sampling time. This study demonstrates the dynamic nature of coastal sandy sediments in terms of microbial diversity, allowing for the formulation of strong ecological hypotheses to explain this phenomenon: strong vertical shifts in nutrient, organic matter, and oxygen availability create a large range of microbial niches, which may support a high turnover of bacterial types in sandy sediments.

Diversity Varying with Oxygen Availability in the Black Sea

The Black Sea, a permanently anoxic basin connected to the ocean through the Bosphorus Sea, has well defined redox gradients and known microhabitats for different metabolic groups of bacteria. C.A. Fuchsman and colleagues (unpublished observations) obtained bacterial pyrotags from four low-oxygen Black Sea water samples: a low-oxygen sample (30 µM oxygen), a sample from the middle of the suboxic zone (2 µM oxygen), a sample from the bottom of the suboxic zone with no detectable oxygen or sulfide, and a sinking particle obtained from the middle of the suboxic zone. The bottom of the suboxic-zone sample (0 µM oxygen) and the particle attached bacterial sample were more diverse than the 30 µM oxygen and 2 µM oxygen samples.

Although all three samples contained low oxygen and no measurable sulfide, only the microbial community structures for the 0 or 2 µM oxygen samples had similar community structures (51% similarity by Bray Curtis) whereas neither resembled the 30 µM oxygen samples (11%). Micro-aerophilic heterotrophs and nitrate reducers dominated the 30 µM and 2 µM oxygen samples. In contrast, the 0 µM oxygen sample and the particle attached bacterial samples were more diverse and contained strikingly different taxonomic groups of bacteria. Enriched populations of *Deferribacter*, ä-Proteobacteria, *Lentisphaera*, å-Proteobacteria and Planctomycetes ocurred in the particle attached fraction. These taxonomic groups of bacteria are not normally identified as part of the particle attached community from oxic waters. Pyrotags for the particle-associated å-Proteobacteria resemble rRNA sequences from epsilon species that oxidise sulfide. The *Deferribacter* species are known to reduce metals including manganese and iron oxides, or nitrate and elemental sulphur. Lentisphaera occur in anaerobic environments but little is known about their metabolism. The ä-Proteobacteria, which include known sulfate reducers, were found in the 0 μM oxygen sample and in the particle attached fraction. Howerver, the OTUs of ä-Proteobacteria differed between the samples, with *Desulfobacteraceae* dominating the 0 μM oxygen sample whereas the particle attached group could not be assigned to a cultured species.

These results showed a clear correlation between the fluxes and depth of the chemical species such as O_2, NO_3^-, NH_4^+, CH_4, MnO_2, H_2S and the inferred metabolisms of the bacterial OTUs. Manganese and sulfate reducers and sulfide oxidisers dominated metabolic groups associated with sinking particles whereas microaerophilic and nitrate reducers dominated the water column. This study provides insights into the importance of the particle attached bacterial communities and points to the potential biases on bacterial diversity estimation when researchers pre-filter samples for microbial diversity studies.

Community Signatures of the North Atlantic Deep Water Masses

Small size suggests that microbes have high dispersal and high immigration rates, leading to a ubiquitous distribution in the marine environment. However, recent studies demonstrate that microbes can have biogeographic distributions corresponding to individual water masses. Distinct salinity, temperature, and nutrient characteristics differentiate several deep water masses separated by thousands of kilometres of thermohaline ocean circulation. Using bacterial pyrotag sequencing, Herndl and colleagues (unpublished observations) tested this hypothesis by determining the biogeography of bacterioplankton communities following the flow of the North Atlantic Deep Water over a stretch of 8,000 km in the North Atlantic. They focused on the distribution of the abundant versus rare phylotypes to decipher whether rare phylotypes exhibit a similar distribution pattern as the abundant phylotypes or whether they occur ubiquitously. If the rare phylotypes represent a seed bank for the few abundant phylotypes, then the community structure of the rare phylotypes should be fairly uniform across water masses.

Cluster analysis showed that abundant bacterial phylotypes clustered according to the water masses. The samples partitioned into one cluster containing bacterial communities from the subsurface zone, two clusters from the mesopelagic waters, three deep water clusters, and one cluster of bacterial communities from the deep Labrador Seawater. Bacterial community composition of deep waters was less similar than samples from subsurface and intermediate waters. Bacterial communities from the same water mass but separated by thousands of kilometres resembled each other more than communities separated by a few hundred metres at individual sites but originating from different water masses.

The clustering of the rare sequences (frequency less than 0.01% within a sample, including the singletons) was similar to the clustering of the abundant sequences (frequency greater than 1% within a sample), albeit with a generally lower percentage of similarity. Proteobacteria constituted more than 85% of the 1,000 most abundant tags and of these, 52% were á-Proteobacteria, mostly composed of the SAR 11 cluster. The bathypelagic zone had the highest proportion of unassigned bacteria, but showed the highest tag richness and evenness compared to overlying waters. ã-Proteobacteria increased with depth, with higher proportions of Chromatiales and Alteromonadales (8.7%) in the bathypelagic zone than in the subsurface zone. The distinct clusters of bacterial communities in specific water masses reflected the presence of unique phylotypes specific to distinct water masses. The variability in the abundance of tags increased with decreasing overall abundance. The most abundant pyrotags exhibited a ubiquitous distribution pattern whereas the representation of low abundance pyrotags seemed to be specific to different water masses.

In summary, the bacterial rare biosphere in the North Atlantic is water-mass-specific and hence not ubiquitously distributed as previously suggested. Thus, in this example, it is likely that the rare biosphere originates from the more abundant and/or more active bacterial community through mutation. The biogeochemical role of the high richness of the rare microbial biosphere remains enigmatic and deserves further investigation.

A Bipolar Distribution of the Most Abundant Bacterial Pyrotags

Among the globally distributed samples sequenced as part of the ICoMM Community Sequencing Project, Polar Regions contributed 56 datasets divided among five projects (ABR, ACB, ASA, CAM, and DAO). The Polar Realm pie chart shows the taxonomic affiliation for the top 20 most abundant tags from these samples. One of the striking results from our comparative study is that the most abundant SAR 11 pyrotag corresponds to the most abundant polar pyrotag. When we map the distribution of the top 20 most abundant tags, we find that all 20 occur in both the Arctic and Southern Oceans. Our non-metric multidimensional scaling analysis confirms the similarity of certain Arctic and Antarctic samples by way of similarity envelopes that encircle datasets that cluster at 80% similarity levels. Although these tags show a "bipolar" distribution using the V6 region as a metric for comparison, we do not know how these populations compare when looking across the entire genome. This question, as well as the source and persistence of bipolar species, awaits the next decade of Cen MICROBIS and Lipid Maps

ICoMM has also considered phenotypic characters ranging from microbial physiology to metabolic capability in its global marine microbial census. Intact polar lipids provide a case in point. Lipids can be a powerful tool for deciphering microbial communities in present and past environments. However, in contrast to genetic data, lipids lack a large database linking identification, chemical structure, and biological or environmental sources. Such a linked database would substantially increase the potential for lipids to be used as a tool for relating environmental microbial communities to cultivated microbes and provide support for phylogenetic relationships. Our approach to constructing a lipid database follows MICROBIS in which a central database/program/website links several databases including a database of microbial organisms, lipid structures (Lipid Maps), and mass spectra.

The Lipid Maps database represents the most extensive lipid library, but it currently targets biomedical applications and lacks a comprehensive collection of marine microbial lipids. Collaboration with Lipid Maps to generate a universal lipid collection that includes marine microbial lipids enabled the deposition of more than 200 marine microbial lipid structures in the Lipid Maps database.

The Lipid Maps database uses a systems biology approach for the categorisation, nomenclature, and chemical representation of lipids (Fahy et al. 2005). The classification scheme of Lipid Maps distributes lipids into eight defined categories that are divided into classes and subclasses. Each lipid in the Lipid Maps library can be retrieved using different search criteria including its lipid identification, classification, systematic name, synonym, and chemical name and structure. Furthermore, Lipid Maps assigns unique numbers to lipid structures comparable to the unique accession numbers within GenBank for genomic sequences. New lipid structures can be continuously submitted to Lipid Maps with a proposed lipid identification and thus can evolve continuously if support within the biogeochemical community is strong.

Typically, mass spectrometry identifies lipids in the laboratory. The mass spectra of lipids are generally very comparable between laboratories and thus well suited for database purposes. Unfortunately, at present there are no publicly available mass spectral databases and only commercial libraries such as those from the National Institute of Standards and Technology exist but do not contain a large number of typical marine microbial lipids. Within ICoMM's database MICROBIS we therefore developed a mass spectrometry library that contains lipid data derived from microbes from modern and ancient environments. This library can run under National Institute of Standards and Technology SEARCH software, which is the most commonly used software to search with mass spectra in mass spectral libraries. At this point we have assembled more than 200 mass spectra of the most common and diagnostic lipids of marine microbes.

MICROBIS has laid the foundation for an integrated database for searching and relating lipid data with other molecular and geospatial data. Once further developed, the MICROBIS website will cross-reference lipidomic, taxonomic, DNA sequence, and geospatial data. Currently, ICoMM is designing a search-engine supported mass spectrometry library wherein cross-referencing between Lipid Maps and the mass spectrometry database will proceed by the LIPID identifications that will also be linked to geospatial data. In the future, MICROBIS will provide a user-friendly interface that will allow searching for taxonomic, DNA sequence, and phylogenetic information, enabling biogeochemists to link lipid data to genomic and geospatial data.

Looking Back in Time with Lipids

ICoMM has also attempted to explore the unknowable: a glimpse at microbial diversity in the past. Dating of evolutionary events within phylogenetic clusters is problematic as it mostly relies on the morphological identification of fossilised remains of microbes, which are usually limited to microbes having inorganic skeletons such as diatoms and coccolithophorids. A new approach is to use fossilised organic molecules. Indeed, fossil DNA occurs in several selected cases (Fish et al. 2002; Coolen et al. 2004), but findings such as these are rare and controversial. In contrast, fossilised lipids commonly occur in sediments of up to 2 billion years old and thus may be, as long as they are diagnostic for certain microbial phylogenetic clusters, suitable to trace the evolutionary history of microbes.

The usefulness of this approach lies in a detailed study of 18S rRNA genes and lipid biomarkers of more than 100 representative marine diatoms (Sinninghe Damsté et al. 2004). This study revealed that several lipid biomarkers are quite specific for hylogenetic clusters within the diatoms. For example, the biosynthesis of so-called highly branched isoprenoid alkenes is restricted to two specific phylogenetic clusters, which independently evolved in the centric and pennate diatoms (Sinninghe Damsté et al. 2004; The molecular record of C_{25} highly branched isoprenoid chemical fossils in a large suite of well-dated marine sediments ad petroleum reveals that the older cluster, comprising rhizosolenoid diatoms, evolved 91.5-1.5 million years ago (Upper Turonian), enabling an unprecedented accurate dating of diatom evolution.

Viewing Microbial Diversity Through a Community Lens

ICoMM's systematic and high-throughput analyses of the sequence variation of rRNA genes have provided a wealth of microbial community sequence data for numerous, poorly understood marine environments. When contextual parameters are recorded together with diversity data, it is now possible to assess the impact of space, time, and complex envronmental gradients on microbial communities, and to quantify interactions among factors. The integration of laboratory-developed microbiological sensors into observing platforms that track changes at high temporal and spatial resolution will enable autonomous observation of changes in marine microbial diversity in the field (Paul et al. 2006). Here we can find answers to as different questions as the following: Why do specific communities flourish in one environment and not in another? Which microbial populations are more successful than others in the competition for energy or space? Which environments host those seed populations that only temporarily dominate communities?

The same tools that have been used in classical community ecology are available for the analysis of changes in microbial community patterns, because ultimately standard sample-by-species matrices can be obtained with any high-throughput method. Our next challenge will be the generation of microbial diversity theories that will allow further comparisons with established ecological theories for macro-organisms or that can be tested across various ecosystems. For instance, recent developments in the study of microbial biogeography may be seen as a prelude to a more dramatic revolution in better understanding microbial communities in their complex environments.

Marine Microbes and Their Roles in a Changing Ocean

The importance of marine microbes to our biosphere cannot be overstated (Box 12.2). Since the microbial census began, several major scientific breakthroughs in microbial diversity and microbial ecology have occurred. Owing to the rapid developments in high-throughput and relatively cost-effective sequencing technologies like massively parallel DNA sequencing, it has become possible to deeply explore microbial (that is, bacterial, archaeal, and eukaryotic) genetic diversity of environmental samples in both qualitative and quantitative ways. Over the past five to ten years, spectacular findings have highlighted new and unexpected roles of microbes in biogeochemical cycling of carbon, nitrogen, sulfur, iron, and many other (trace) elements owing to interdisciplinary research based on the integration of sequencing, membrane lipid research, and isotope techniques.

Fascinating examples of new and important microbial shunts in biogeochemical cycles include the following: the existence of anaerobic oxidation of methane by bacterial–archaeal consortia oxidising methane with sulfate operating in (sub)oxic environments in the ocean and on land (Knittel & Boetius 2009 and the literature cited therein); the anaerobic oxidation of methane by bacteria oxidising methane with nitrate (Raghoebarsing et al. 2006); anaerobic ammonium oxidation (Anammox), whereby anammox bacteria use specific membrane components to oxidise ammonia with nitrite to form nitrogen gas that escapes from the oceans (Strous et al. 1999; Sinninghe Damste et al. 2002; Kuypers et al. 2003); the discovery that some Crenarchaea use ammonia as an energy source (Konneke et al. 2005); crenarchaeotal CO_2 fixation (Wuchter et al. 2003); and the incredible phenotypic adaptation of the largest known bacterial cells on Earth, the giant sulfide oxidising bacteria, to their environment (Gallardo 1977; Teske & Nelson 2006).

Marine Microbial Diversity and Abundance Hghlights:

- The number of bacteria in the open ocean exceeds 10^{29} cells and microbes in total contribute as much as 90% of the biomass in the ocean.
- Microbes may be more than 100 times more diverse than plants and animals.

- A single litre of seawater can represent approximately 20,000 "species" of bacteria.
- A gram of sand can contain between 5,000 and 19,000 "species" of bacteria.
- Archaeal cell numbers can rival those of bacteria in the ocean but their diversity is 10% that of bacteria.
- Protist diversity rivals that of Archaea in some parts of the ocean.
- Most of marine microbial diversity is represented by low abundance populations.
- Each metazoan may have its own unique microbiome population structure.
- Different water masses possess signature microbial community structures.
- Some microbes are everywhere: ICoMM's most abundant type of bacterial pyrotag matched sequences from the SAR11 marine bacterial group which accounts for 25% of the biomass and 50% of the cell abundance in the pelagic ocean.
- Lipids allow us to look back in time at ancient microbial populations and delve into the unknown.

These and many other examples (Giovannoni & Stingl 2005; Karl 2007; Azam & Malfatti 2007; Bowler et al. 2009; DeLong 2009; Fuhrman 2009) clearly indicate that marine microbial biogeochemical cycling of elements is even more important than traditionally thought and that microbes dominate these cycles in many known and recently discovered ways. This increased recognition of microbial importance in biogeochemical cycling of elements combined with the discovery of vast microbial diversity, the discovery of the rare biosphere, and the dominance of just a few microbial taxa in environmental settings, makes us aware of the crucial importance of microbes in climate and climate change.

In ICoMM we are aware that ongoing human-induced global climate change will and probably already has impacted microbial diversity. Owing to rising seawater temperatures, ocean acidification, and salinity changes, dominant marine microbial taxa may become dormant and completely unknown taxa present in the environment but extremely rare, may become dominant. Because we have little idea about which marine microbial taxa will become dormant and which will become dominant, predictions of changes in biogeochemical cycles are very difficult to make. Thereby predictions of not whether but rather how the climate will change and how to ameliorate such change present even greater challenges to scientists and policy-makers.

The above can be illustrated by the following. The far greater part of N_2 fixation in the marine environment is currently performed by just a very few bacteria, *Trichodesmium* and an uncultivated 'Group A' putative unicellular cyanobacterium lacking oxygenic photosystem II, whereas the endosymbiotic cyanobacterium *Richelia intracellularis*, as well as other endo and ectosymbiotic N_2-fixing cyanobacteria, are less important globally (Arrigo 2005). Very preliminary laboratory experiments with artificially acidified seawater indicate that N_2-fixing bacteria react very strongly to lowered pH values, thereby changing the rate and nature of nitrogen and carbon fixation (Hutchins et al. 2007). This will impact the complete marine nitrogen and carbon cycles and thereby other biogeochemical cycles (for example, Crenarchaeota use ammonium to fix carbon; phosphate may become the omnipresent limiting nutrient). Similar experiments with the major carbon-fixing organisms in oligotrophic water, *Prochlorococcus* and *Synechococcus*, also indicated that acidification and temperature rise will severely affect their metabolism, leading to changes in the rate of CO_2 fixation (Fu et al. 2007).

There are many factors that add to the uncertainty in the estimation of the pathways and the scale of the interaction of marine microbe communities with anthropogenically driven global climate change. Second only to human impacts on the environment, the interplay between environmental parameters and shifts in marine microbial diversity will dominate the course of climate change. The paucity of research on underlying mechanisms severely constrains the ability of policy-makers to make informed decisions about mitigating strategies. Trying to understand microbial diversity and functioning in biogeochemical and nutrient cycling is of major importance for future research on a worldwide scale

Outlook

Looking forward to the next decade of microbial census research, we see many opportunities and challenges. As massively parallel DNA sequencing brings an unprecedented volume of data, it also brings challenges in analysing these data. In this chapter we have chosen to highlight the general findings of our efforts so far, but the necessary computer algorithms and models required to bring us closer to more robust estimates of microbial diversity are still being developed and the required computational power still being sought. Improving the taxonomy attached to pyrotags is another area that will need attention in the future. Much of this will likely come though improved annotations of the vast amount of full-length or nearly full-length sequences presently housed in public databases, as well as next-generation sequencing providing much longer reads that will enable better taxonomic assignment. Yet, we find that even when definitive taxonomic assignment is not possible, the approach is still very powerful for comparisons of assemblage composition and diversity. The technique reveals substantial diversity undetected by previously used techniques.

Even ICoMM formalin-preserved samples belonging to the JointGlobal Ocean Flux Study have been successfully sequenced, unveiling the potential of pyrosequencing to become a powerful tool for paleobiological and paleoenvironmental studies. The massive amount of data from pyrosequencing also enables predictions of functions for many OTUs. In short, large-scale patterns could emerge not only of phylogenetic diversity but also functional diversity along geochemical gradients. As the 454-pyrotag based technology does not distinguish between dead and living micrbial taxa, it could be complemented by other techniques for assessing the level of viability within a sample such as starting with RNA samples and reverse transcribing it to reveal the most active populations.

Deciphering ecological signals linked to the definition of rare or abundant OTUs at different taxonomic levels is crucial. This definition will influence how diversity patterns are measured. Furthermore, this new paradigm will aid researchers to better understand marine microbial communities and their impact on planetary biogeochemical cycles.

Marine Animal Populations: A New Look Back in Time

Future endeavours must pay closer attention to the temporal dimension of changes in microbial community structures. As sequencing costs decline, it will be possible to monitor microbial populations and their changes over time in appropriate marine environments on different timescales from minutes to centuries. Developing such monitoring strategies through existing observing systems, time-series stations (for example Bermuda Atlantic Time-Series, Hawaii Ocean Time-Series), and Long Term Ecological Research Sites, we might be better able to predict changes in microbial populations as a consequence of natural and anthropogenic climate change. The ICoMM team recommends that this kind of monitoring approach may thus help substantially to improve our ability to predict climate change, harmful algal blooms, and ultimately our own impact on Marine Animal Populations: A New Look Back in Time.

Introduction

Since around 1980, marine-capture fisheries have stagnated at around 90 million tonnes per year, despite massive technological investments and the opening up of distant and deep waters in the Southern hemisphere. The oceans will simply not yield more. In fact catches are of increasingly smaller fish of less economic value and total returns on investments are dwindling. On a global scale, capture fisheries are doomed to be of less importance as a source of protein to a growing human population, while the fishing pressure remains extremely high. There is no sign that the rise of aquaculture in recent decades has eased the pressure on wild resources.

The fisheries crisis is part of a general health alert for the oceans. Marine habitats are under severe pressure as a side effect of trawling and directly by dredging, harbour development, the concretisation of large stretches of coastline, and especially from eutrophication caused by both agriculture and aquaculture (Lotze & Worm 2009).

But what is the scale of change? What used to be in the sea before humans began impacting marine ecosystems and habitats? What are the major long-term effects of human extractions of marine life? Are the impacts of recent or ancient origin? In other words what are the baselines against which we may evaluate some of the findings of the Census of Marine Life field projects by 2010? Can we talk with confidence about the history of the sea, can we gauge how much has changed – and with what consequences to us humans? This was the grand challenge that was put to the scientific community some ten years ago when the Census endorsed the History of Marine Animal Populations (HMAP) Project to assess and explain the history of diversity, distribution, and abundance of marine life (Box 1.1).

Regional and Species Focus of HMAP

HMAP is a collaborative effort by some 100 researchers around the globe participating in several region- or species-specific research teams. Twelve are based on marine areas, as follows: southeast Australian Shelf; New Zealand Shelf; Caribbean Sea; Gulf of Maine; Newfoundland and Grand Banks; Baltic Sea; North Sea; Mediterranean Sea; Black Sea; White and Barents Seas; southwest African Shelf; and the biodiversity of nearshore waters. Three case studies focus on the following species: whales, cod, and mollusks and one on Northern European fish bone assemblages. In addition, several smaller case studies have been undertaken in areas such as the Philippine Seas, the Wadden Sea, and the seas of Indonesia and northern Australia.

Although the history of marine animal populations has long been one of the great unknowns, recent advances in scientific and historical methodology and new applications of existing methodology have enabled the HMAP teams to expand the realm of the known and the knowable (Holm 2002).

The analytical framework of HMAP embraces two basic premises, one concerning data, one concerning methodology. First, much of what we can know about the history of the oceans will be in the "human edges" of the ocean, those in the near shore and coastal zone. This is where humans most directly interacted with the sea in the past and therefore most historical records relate to these activities. However, in both the human edges and in the central oceanic waters there have been extensive fisheries for larger organisms, and the value of the organisms encouraged the creation and maintenance of archival material. As HMAP has evolved, new and unexpected data sources have been discovered, and we know now that vast repositories are still untapped.

Second, historical analysis must combine with ecological analysis in a truly interdisciplinary way. New insights are due to the introduction of established marine science methodology to historical data, notably standardising fishing effort (catch per unit effort), biodiversity counts of historical fisheries (Lotze et al. 2005), statistical modelling of historical data (Klaer 2005; Rosenberg et al. 2005), etc. Perhaps the most surprising results have come simply from the data-mining effort in itself, which has revealed a wealth of documentation for historical fisheries previously neglected by historians.

Examples of this are of catch records spanning two to four centuries (Holm & Bager 2001; Starkey & Haines 2001; Lajus et al. 2005; B. Poulsen 2010). HMAP has provided inspiration to glean important information from surprising and sometimes unlikely sources such as restaurant menus (Jones 2008) and snapshots of sports fishermen's catches (McClenachan 2009). Archaeological techniques have been deployed in conjunction with historical methods and stable isotope analysis to explore the character and composition of fish catches during early medieval times (Barrett et al. 2008), and many more unconventional approaches could be cited.

In many ways the complicated interplay between man and nature calls for a new type of historical research. Science is a challenge to historians who have had little statistics, not to speak of modelling, as part of their training. Historical source-criticism is a challenge to scientists who are used to hard data. Although academic history through the 1990s concentrated on narrative and deconstructing skills, environmental history also demands command of both statistical and scientific methods.

The need for historians and scientists to work together is not uncontroversial. Some historians assert that history would carry no lessons for the future as events are never repeated in exactly the same form. Some scientists doubt the validity of historical data that are by definition "dirty data" in the sense that they are relics of events, not signals of a recurrent phenomenon, or experiment, established in a controlled environment such as a laboratory. In the early part of the Census some skeptics doubted the role of environmental history in this mega-science program. Would such a program not by default perpetuate the divide between science and the humanities? Indeed, as one critic put it, would the marriage of history and science not lead to scientists simply appropriating data for their own use (Van Sittert 2005)?

HMAP is founded on the belief that the divide between history and science needs to be bridged. History will never repeat itself but like the child learns to walk based on experience so does society base decisions and preferences on past experience. The historian may indeed detect trends and patterns of behaviour behind diverse and unique events. Emphatic statements on the validity of and need for the HMAP approach have been made by some historians (Anderson 2006; Bolster 2006, 2008). Conversely, if we reduce science to controlled experiments we would never understand the fundamental principles of natural selection. More urgently, contemporary concerns about global climate change, biodiversity, and scarcity of resources are based on perceived changes of nature and availability of natural resources. Therefore, the history of nature itself – and the dependency and impact of human society on nature – has become a prime social, economic, and political concern, and scientists and historians need to address these very real issues, or decisions will be based on assumptions.

Environmental historians do not have to become biologists, nor do biologists need to become historians. However, we do need to understand enough of each other's language to exchange information and insight. Our experience of dialogue across the current divide of humanities and science has led to the emergence of the new scientific community of marine environmental history and historical marine ecology

A total of 205 books and papers have been published up to September 2009 and the HMAP database holds approximately 350,000 records, with some 80% available through OBIS. By late 2010, it is anticipated that up to 1,000,000 records will be available on the HMAP website. With such a massive output it is obvious that any overview of major findings will be highly selective. In the following, we shall establish first the state of knowledge before the beginning of the Census in 2000, then focus on some of the highlights from the HMAP case studies. By way of conclusion, the chapter closes with observations on what we do not know, how we may get to know it, and why some questions will remain unanswerable.

The Background

Marine ecology was born as a scientific discipline by the late nineteenth century and derived often from a strong interest in the fisheries (Smith 1994). The question of human impact on marine life was central not only from the perspective of economic interest (for example where are the fish and how do we catch them?) but from the perspective of human impact (for example what is the effect of extracting thousands of tonnes of fish and what damage to the seabed may be caused by certain fishing technologies?).

The central question of the possibility of overfishing was raised at the World Fisheries Exhibition in London in 1884 and drew two opposing answers. One came from one of the leading scientific figures of the day, Thomas Henry Huxley, who concluded that "probably all the great fisheries are inexhaustible; that is to say that nothing we do seriously affects the number of fish" (Huxley 1883). A more conservative note was struck by Ray Lankester, a young professor of zoology, that "the thousands of apparently superfluous young produced by fishes are not really superfluous, but have a perfectly definite place in the complex interactions of the living beings within their area" (Lankester 1890). To the credit of both men and to the academic community at the time the question of the possibility of harmful overfishing was put to the test. A rigorous series of trawls were undertaken in Scottish waters and were at first understood to support Huxley's view. In 1900, however, the tests were re-analysed and further data from observations of commercial operations out of Grimsby were scrutinised. The conclusion by Walter Garstang was clear and had far-reaching implications: "the rate at which sea fishes reproduce and grow is no longer sufficient to enable them to keep pace with the increasing rate of capture. In other words, the bottom fisheries are undergoing a process of exhaustion" (Garstang 1900; cf. Smith 1994, pp. 106–108).

This fundamental observation is at the heart of the question of human interaction with the oceans. Garstang established beyond scientific doubt that extractions might have an impact. Through the twentieth century, fisheries science concentrated on identifying optimal sustainable yields that would not extract more from the sea than marine life would be able to replenish. By the second half of the century, fisheries science had become highly sophisticated, equipped with research ships and advanced computer models. Scientific organisations like ICES, the International Council for the Exploration of the Sea, established in 1902 for the North Atlantic (Rozwadowski 2002), and a plethora of similar organisations for other ocean realms and migratory species, struggled to get both the science right and deliver management advice. Characteristically, fisheries studies were often based on very short time-series, although scientists were aware of long-term changes.

The centennial variability of the Swedish Bohuslen herring fisheries provided a textbook example that fisheries may change dramatically over the long term. Nevertheless, perhaps because of the strong link with policy advice, the focus of cutting edge science tended to be on recent data often obtained with new equipment, which by the very fact oblitreated longer-term perceptions. Data observations over the long term were often discontinued for financial reasons. Few observation series are maintained today that span more than a few decades, the best-known of which is the Continuous Plankton Data Recorder survey, which has been maintained for the North Atlantic and North Sea since 1931 (Continuous Plankton Recorder 2009).

Marine science separated from fisheries science through the twentieth century as scientists developed the concept of ecology as a study of biodiversity, food webs, and biological processes and functions as a separate line of inquiry. To ecologists the ultimate question is not what is in nature for us, the humans, but how do we understand nature on its own, with the humans left out. Interest focused on biodiversity, the awesome richness of nature, and the exhilaration of understanding intricate and ingenious life-forms. By the 1960s ecologists did realise that ecosystems rarely remain steady for long, and "fluctuations lie in the very essence of the ecosystems and of every one of the populations" (Margalef 1960 cited in Smith 1994, p. 33). Marine ecologists, however, perceived little or no need for history, with the exception of a few studies of correlations between contemporary and historical observations of animal populations and key environmental variables (Cushing 1982; Alheit & Hagen 1997; Southward 1995). Things were about to change, however, as demonstrated in a programmatic statement on the need to determine the historic structure of exploited ecosystems (Pitcher & Pauly 1998).

In a seminal study of the Caribbean ecosystem, Jeremy Jackson criticised ecologists for assuming that the natural or original condition is equal to the first scientific description of a phenomenon (Jackson 1997). Jackson turned to a concept developed a few years earlier by a fisheries scientist, Daniel Pauly, for a diagnosis of the problem, which was termed the shifting baseline syndrome (Pauly 1995). Pauly observed that equilibrium or steady-state models are based on a given dataset, often established by scientists within the past generation. However, what happens to equilibrium if older data are introduced? We cannot know from recent information the extent of the losses that have happened.

Jeremy Jackson, himself an American ecologist, son of a historian, used the British Empire trade statistics of the eighteenth century to learn of the trade in turtles from the Caribbean. When working out the numbers hundreds of thousands of turtles killed in a single year – he realised that the ecosystem of the Caribbean would have looked very different to what conservation biologists supposed based on information from the past couple of decades (Jackson 1997). The lesson to ecologists of Jackson's historical analysis of Caribbean coral reefs was that textbook descriptions of reef ecosystems were limited by the fact that the systematic description by modern biology only began in the 1950s. Jackson put the case squarely to the ecologists: they needed to turn to historical sources and rediscover the world.

Another influential development in reinstating the historical dimension in science was the development of paleoecology and archaeoichthyology in the past 30–40 years. The preservation of fish scales in anoxic bottom sediments off the coast of California provided scientists the opportunity to reconstruct 1,600 years of pelagic abundances (Soutar 1967; Baumgartner et al. 1992; Francis et al. 2001). The field of paleozoology provided one of the first clear examples of scientists working across the cultural divides of historical and ecological analysis. Analysis of fish remains from archaeological sites provided a possible avenue to understanding biodiversity distribution and abundance. In the 1960s the Swedish scientist Höglund analysed fish bones excavated from eighteenth century production sites for train oil and found that the Bohuslen herring was spent (namely post-spawning) herring from the sub-population of the North Sea Buchan herring (Höglund 1972). In the 1990s studies clearly demonstrated the potential of bringing the different lines of inquiry together (Muniz 1996; Enghoff 1999).

What about the historians? Environmental history has been a growth field in the USA since the 1970s and a little later in Europe, Asia, and Australia, and indeed, despite institutional problems, also in South America and Africa. However, the focus by leading American environmental historians was strongly on human agency and perception whereas ecological factors were rarely allowed to play an explanatory role. On top of that, the discipline developed out of a strongly narrative and qualitative approach to history that had little rapport with the quantitative approach of ecologists. The focus was very much on frontier cultures of the prairies, bushlands, savannahs, and steppes, whereas the oceans were strangely disregarded. Maritime historians on the other hand were firmly embedded in economic and social history with a preoccupation for naval and shipping matters and had little regard for environmental issues.

The few fisheries historians often found their subject of marginal interest to mainstream historians and a bit fuzzy as the ecological context of fishing could not be disregarded but on the other hand was little understood. The few substantial overviews of fisheries published generally adopted a national, regional, or port perspective whereas environmental considerations were accidental at best. It was only as late as 1995 that the North Atlantic Fisheries History Association was established, but even then few papers dealt with the impact of harvesting on the seas (Holm & Starkey 1995–1999).

Signs were in the air, however, that things were about to change. In the North Atlantic, Holm & Starkey (1998) reported the results of a workshop titled "Fishing Matters" that brought together historians, social scientists, biologists, oceanographers, and fisheries managers to examine multidisciplinary approaches to understanding the past and current scale and character of the fisheries. In the North Pacific, Pauly et al. (1998b) similarly documented the results of a workshop aimed at mathematically reconstructing the state of the Strait of Georgia, off Vancouver Island. Participants were even more varied, and the focus was broader in attempting "to provide a vision for rebuilding the Strait's once abundan The HMAP Projects

Such was the state of play when a preparatory workshop of the Census in 1998 called attention to the need of a historical backdrop – a baseline – to observations of ocean life (Anon 1998). The challenges were apparent: there was no shared or agreed set of methodologies and not even agreement as to which research questions needed to be raised. Before the project started the first step was therefore to bring together a workshop in February 2000 to identify the hypotheses that could be tested against historical data, to identify the various sources of data, and the methodologies that might yield plausible answers.

The workshop agreed that historical data were only sporadically available and that there was an urgent need to build consistent time-series of extractions and fishing effort for at least the best documented operations such as whaling and large commercial fisheries. Participants identified 10 hypotheses to direct work in the early years of the project. The focus was first of all to investigate the proposition that validated historical, archaeological, and paleoecological records can be used to gauge long-term change in the abundance, spatial distribution, and/or diversity of marine animal populations. Secondly we wanted to identify the environmental and human forces that might condition fish mortality. Thirdly, we wanted to understand better the drivers of these forces themselves, be they related to geophysical or human activity.

In May 2000 a Steering Group of historians and marine scientists was charged by the Census' Scientific Steering Committee to lead a global inquiry into the history of marine animal populations. A series of regional projects was proposed while we set up annual training workshops through the summers of 2001–2003, well knowing that as nobody had ever received academic training as marine historical ecologists or marine environmental historians, there was a need to train a new generation of two dozen young researchers to understand enough of several disciplines. As the project grew, the Oceans Past conferences of 2005 and 2009 attracted more than 100 researchers while many more worked in the field.

The identification of a viable project was not just a question of a top-down process. Although the Steering Group wanted to get projects started in Japan and in the American Pacific, we were confronted with the reality of needing to find like-minded people who would undertake not only individual work but also lead a team for several years guided by an overarching research program. We were not always successful, but all projects that were begun proved viable. Although some were discontinued as the research was completed, other projects developed new agendas. A renewed focus on the evidence of archaeology brought new people and projects forward. By 2007 the focus shifted from data collecting to synthesis, both within projects and across projects. New collaboration with other Census projects emerged, in particular with Natural Geography in Shore Areas (NaGISA) for the History of the Near Shore project, which focused on providing historical data as baseline studies for ongoing fieldwork. In the following, we highlight selected research findings addressing two of the initial simple questions: what is the scale of change, and are changes of recent or ancient origin?

Mediterranean Sea and Black Sea

The Mediterranean and Black Seas are among the earliest heavily fished marine ecosystems in the world. Fish as a source of food was more important than meat in the ancient Mediterranean cultures. Along the Nile, settlements with huge amounts of fish bones have been identified. Hundreds of full-time fishers were employed by the Lagash temple in Sumer around 2400 B.C. The fish was dried, salted, and stored. Babylonian sources from around 1750 B.C. show the importance of fishing. Greek merchants conducted an extensive fish trade from the Black Sea and the Russian rivers to the Gre However, the problem with assessing the impact of fisheries on ecosystems is that ancient records are rarely quantifiable and often we are not able to identify the fish species mentioned. Even worse, until quite recently, historians have assumed that the ancient fisheries were of minimal importance, technology was simple, and nets were cast close to the shoreline.

A full reversal of this perception was only achieved as a result of an analysis of the evidence matched by an understanding of modern impact studies of pre-industrial fisheries technology. The Graeco-Roman world had seagoing vessels for hook-and-line as well as net fisheries. Ancient technology was neither ineffective nor unproductive, and indeed produced such large catches that the limiting factor was preservation and storage (Bekker-Nielsen 2005). The main fisheries for bluefin tuna (*Thunnus thynnus*), mackerel (*Scomber scombrus*), and other pelagic species took place in narrow straits such as the Strait of Gibraltar, Sardinia, Sicily, and Crimea in the Black Sea (Curtis 2005; Gertwagen 2008).

One solution to the problem of conserving the fish was to dry and salt the fish, which was done extensively and accounted for much of the Greek imports from the Black Sea. The most spectacular solution was, however, the reduction of fish to fish sauce, garum, essentially by throwing the catch into large containers to allow a fermenting process to result in a liquid that was then traded all around the Roman world to add flavour to the Roman cuisine.

The large installations are especially found by the shores of the western Mediterranean and the Black Sea. They were probably privately owned commercial operations for export, and regularly had containers of several hundred cubic metres. The largest installation in present day Mauretania had a capacity of over 1,000 cubic metres (Curtis 2005; Trakadas 2005).

As yet, there is no way to establish the quantities of catch, although evidently they will have been significant. One assessment of the distinctive amphora vessels for the oil, wine, and garum trades established that wine accounted for about 62% of relative volumes, whereas oil made up about 28%, and 10% contained garum. Fish sauce was sold all over the Roman Empire and was an essential part of the Roman dish, part of what made up Roman culture (Ejstrud 2005). There is no doubt that extractions will have been huge, and much will be learnt in coming years as this research continues.

Documentary records are especially rich for the Venetian lagoon and the Northern Adriatic Sea. Preliminary studies show that the marine system has been modified dramatically by human interventions since the medieval period. An ongoing project aims to reconstruct the dynamics of marine animal population in the Venetian Lagoon and in the Northern Adriatic Sea from the twelfth century up to the twenty-first century from historical and scientific sources (Gertwagen et al. 2008). Finally, the Catalan Sea has been studied carefully and data for twentieth-century fisheries have been made available for further study.

North Sea and Wadden Sea

The North Sea is another heavily fished and depleted marine system. The Mesolithic period about 6,000 years ago experienced a warm climate, which seems to have been conducive to extensive fisheries all over the Northern hemisphere. Many basic technologies for the fisheries were already developed by this time such as trap gear and fishing by hook-and-line from a boat. With domestication of animals and development of agriculture in the Neolithic period about 5,000 B.P., hunting and fishing became less important and settlements were no longer related to the seashore, and fishing seems to have been of minor importance through the Bronze and Iron Ages of Northern Europe. Rivers will have brought nutrition to the North Sea from the rich agricultural lands of Northern Europe already by the Bronze Age when major deforestation took place and increased the productivity of the sea (Enghoff 2000; Beusekom 2005).

Our knowledge of ancient fisheries is still deficient due to the lack of sieving of archaeological finds for small and easily overlooked fish bones. However, thanks to a thorough review of archaeological reports of dozens of medieval settlements we now know that the period ca. 950–1050 saw a major rise in fish consumption around the North Sea (Barrett et al. 2004, 2008). Early medieval sites are dominated by freshwater and migratory species such as eel and salmon, whereas later settlements reveal a widespread consumption of marine species such as herring (*Clupea harengus*), cod (*Gadus morhua*), hake (*Merluccius*), saithe (*Pollachius virens*), and ling (*Molva molva*). The "fish event" of the eleventh century reflected major economic and technological changes in coastal settlements and technologies, and formed the basis of dietary preferences that were to last into the seventeenth century.

In particular, the evidence of traded cod, "stock fish", which begins to show up in Northern European towns by the middle of the eleventh century, is clear evidence of the rise of commercial fisheries. Barrett's group combines an osteological study of fish bones with analysis of their stable isotope signatures. The project has now identified traded cod in medieval settlements from Norway, England, Belgium, Germany, Denmark, Sweden, Poland, and Estonia. The evidence also supports a hypothesis that seagoing vessels were in wide use by the thirteenth century catching fishes at depths of 100–400 m such as ling. Commercial fisheries were well established by the high middle ages to feed a European population that had developed religious practices of fasting and abstinence of red meat in favour of fish on certain weekdays and through the 40 weekdays of Lent (Hoffmann 2004).

The first estimate of total removals of one species from the North and Baltic Seas comes from the sixteenth-century Danish inshore fisheries for herring in Scania and Bohuslen. Annual catches regularly reached a level of 35,000 tonnes (Holm 1999, 2003). By the late sixteenth century, the Dutch had taken the lead in Northern

European herring fisheries with seagoing buysen, which harvested the rich schools off the coasts of Scotland and the Orkneys.

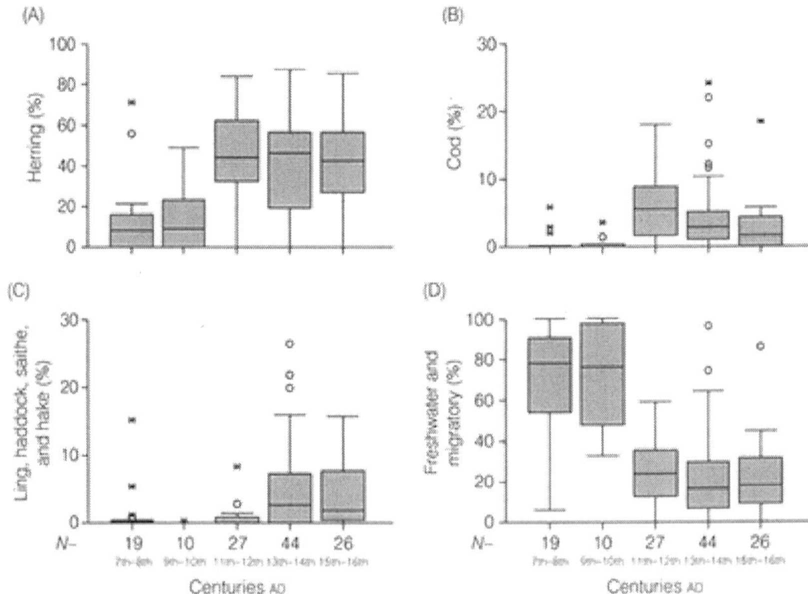

Figure: *Fish bones project: "Pristine" North Sea impacted ca. 950–1050, freshwater to marine species. Source: Barrett et al. (2004). Reproduced with permission.*

They landed catches of 60,000–75,000 tonnes every year in the first quarter of the seventeenth century, and total removals with English, Scottish, and Norwegian landings amounted to upwards of 100,000 tonnes. Catches declined to about half by 1700, and only increased to about 200,000 tonnes in the late eighteenth century owing to Swedish and Scottish progress (B. Poulsen 2008).

By 1870 total removals reached a level of 300,000 tonnes, which equals the recommended Total Allowable Catch for 2007 for herring in the North Sea (ICES 2006). In the twentieth century, total catches repeatedly amounted to well over a million tonnes annually, causing collapses of herring stocks and the closure of fisheries for one or two decades to allow populations to rebuild.

This evidence demonstrates how fishermen in the age before steam and trawl were able to remove large quantities of biomass from the sea. The technologies of wind power and driftnets were practically unchanged in the Dutch fisheries from the seventeenth to the nineteenth centuries. There are indications that removals even at the much lower level than that recommended by modern standards had an effect on abundance. One study standardised the fishing power of North Sea herring fishing vessels across the technological divide from sail to motor-powered vessels from the sixteenth to the twentieth centuries. Even by a conservative estimate the returns of catch per unit effort indicated that stock abundance was ten times higher in the 1600s than in the 1950s, and already by the 1800s, well before the big technological change, it had dropped to 50–60% of the level of the 1600s (B. Poulsen 2008). The effects of early removals may therefore have been larger than we would have assumed.

The catches of two other commercially important species, ling (*Molva molva*) and cod (*Gadus morhua*), were abundant in the nineteenth century whereas the stocks showed signs of depletion by World War I. Detailed historical data are available from the Swedish fishery in the north-eastern North Sea and Skagerrak, which make up about one-sixth of the entire North Sea. Minimum total biomass of cod in 1872 was estimated at about 47,000 tonnes for this portion of the North Sea, but it may have been much higher, whereas the total biomass of ling was estimated at a total of 48,000 tonnes.

These were very healthy stocks if the levels are compared with the modern biomass estimate for cod of 46,000 tonnes for the entire North Sea, Skagerrak, and Eastern Channel, whereas for ling no modern biomass estimate is available as the species is caught too infrequently. The cod population is today considered severely depleted throughout the North Sea, and the ling population may be considered commercially extinct from the region that once produced the major catches (R.T. Poulsen et al. 2007).

Ecosystem theory emphasizes the importance of top predators for the entire food web. Top predators play a controlling and balancing role for the abundance of other species further down the food chain, and an abundance of top predators is a sure sign of biodiversity (Baum & Worm 2008; Heithaus et al. 2008). Human hunting tends to focus on top predators as the big fish are of highest commercial value. When we take out the largest specimens, we remove one of the controls on the ecosystem. The mature fish are also highly important for the reproduction of the population as their eggs have been shown to be healthier and more plentiful than the spawn of younger and smaller specimens (O'Brien 1999). Because the fish continues to grow through its entire life, a decline in the length of specimen caught is a clear indication that the fishery is changing the age structure and viability of the stock. Analysis has shown that whereas the average length of northeastern North Sea ling in the mid- to late nineteenth century was about 1.5 m, it had decreased to about 1.2 m by World War I, and ling caught today is less than 1 m on average (R.T. Poulsen et al. 2007). A century ago, cod landed from the North Sea was usually 1–1.5 m long whereas today it is only about 50 cm.

This means that although cod used to live to an age of 8 or 10 years, today it is caught at less than three years of age; for example in 2007, 87% of the catch in numbers were aged two years or younger. As cod only spawns at the age of three years, this fishing pattern is inhibiting the population from maintaining itself and delaying recovery (ICES 2008). The bluefin tuna (*Thunnus thynnus*) generally escaped human hunting activity until the twentieth century owing to its rapidity and superior strength, which made its capture difficult. By the 1920s superior hook-and-line technology was available and brought tuna within the reach of fishermen. Even more importantly, harpoon guns and purse-seining methods, eventually implemented with hydraulic winches, were developed in the 1930s and rapidly increased catches to thousands of individuals per year.

By 1960, however, tuna catches were falling and ceased to be of commercial importance after the mid 1960s. Climate change and prey abundance seem unlikely causes for the sudden decline, and it seems now likely that the commercial extinction of bluefin tuna from the North Sea was caused by the heavy onslaught by humans in the mid-twentieth century (MacKenzie & Myers 2007).

In the southern North Sea, the haddock (*Melanogrammus aeglefinus*) fishery was of substantial size in the sixteenth and first half of the seventeenth centuries. The fishery declined in the later seventeenth into the eighteenth century, but by the 1770s the fishery was on the increase again. We have evidence of an abundant haddock fishery by German and Danish hand liners in the German Bight and along the Jutland coast in the late eighteenth century and first half of the nineteenth century. Statistics show substantial catches by 1875 declining rapidly in the last quarter of the century to nil around 1910.

It would seem that the southern North Sea haddock stocks were rendered commercially extinct by the intensive German and Fanø-Hjerting fisheries of the late nineteenth century. Today, haddock is prevalent mainly in the northernmost part of the North Sea and in the Skagerrak (Holm 2005), whereas its former widespread presence in the southern part of the North Sea was not generally recognised by marine science until its regional history was revealed.

Major changes to the inshore habitats of the North Sea and thus to marine wildlife occurred in the Middle Ages. Hunting and fishing took its toll on the rich wildlife of the inshore areas of the Wadden Sea, a large intertidal zone off the coasts of the Netherlands, Germany, and Denmark. Dikes, traps, and other inshore coastal uses changed the wide mud flats. By the late nineteenth century industrial and chemical pollution began to build up in the sea. However, the major change to the ecosystem is likely to have come from direct effects of removals of animals by fishing and hunting (Beusekom 2005).

Some marine species have been extirpated from the Wadden Sea such as pelicans (Pelicanus crispus), which disappeared about 2,000 years ago (Prummel & Heinrich 2005), the Atlantic gray whale (Escherichtius robustus), which went extinct not only from the nearshore habitats of the North Sea but as a species sometime in the late medieval period (Mead & Mitchell 1984), and the great auk which disappeared from the North Sea by the medieval period before extinction from the North Atlantic by the nineteenth century (Meldgaard 1988).

Several species have been so much reduced in numbers that they are considered regionally extinct or at least so rare that they have lost their ecosystem importance, and their previous commercial importance to the human economy. Sturgeon was previously caught in vast quantities and marketed in the hundreds, for instance at the Hamburg fish auction. By 1900, however, the fishery declined rapidly both because of river and inshore pollution and fisheries. As late as the 1930s sturgeon was still caught regularly in the northern Danish part of the Wadden Sea but is now extremely rare (Holm 2005).

A general survey of extirpations in the Wadden Sea concluded that major impacts occurred by the turn of the twentieth century, well before the introduction of modern industrial fishing technologies to this region. The major causes for species decline and indeed extirpations were associated with removals and habitat destruction whereas factors such as pollution, eutrophication, and climate change have been late and minor factors so far (Lotze Baltic Sea One of the early research questions of HMAP was posed by fisheries scientists about Baltic cod (MacKenzie et al. 2002). In the absence of historical records before 1966, they wondered if the record high cod stock in the Baltic Sea in the late 1970s to early 1980s was a unique occurrence or likely to occur at regular intervals. The question was unequivocally answered by the work of the Baltic team.

Through the recovery of historical data back to 1925 we know now that abundant cod stocks corresponded to a favourable combination of four key drivers in the late 1970s: incursions of saline water to the brackish Baltic and hydrographic conditions allowing successful reproduction, low marine mammal predation, high productivity environment fuelled by nutrient loading, and reduced fishing pressure. A similar situation did not occur at any other time in the twentieth century. The cod biomass in the 1920s–1940s was likely restricted by high abundance of marine mammals and low ecosystem productivity; and in the 1950s–1960s by high fishing pressure. Periods of deteriorated hydrographic conditions occurred throughout the twentieth century and were most pronounced in the past 20 years, thereby restricting cod recruitment (Eero et al. 2008).

Today, cod rarely ventures into the very brackish northern Baltic waters between Stockholm and the Gulf of Riga. In the late sixteenth and early seventeenth centuries the presence of a large cod fishery off southwest Finland indicates that cod abundance must have been very large. The abundance is all the more remarkable because the population of top predators such as seals would have been much larger than today (MacKenzie & Myers 2007). Climate clearly impacted fish distribution but there are some surprises which underline that some fish are quite resilient to change. Archaeological evidence of fish fauna in the Atlantic warm period (ca. 7000–3900 B.C.) shows many fish species in waters around Denmark that we would today expect to find in warmer waters. Indeed, comparison with contemporary data from surveys and commercial landings shows that many of these species are now re-appearing as temperatures rise. However, cod was very abundant in the Stone Age, even though temperatures were 2–4 °C warmer than late twentieth century temperatures. This finding suggests that commercially important cod populations can be maintained in the North Sea–Baltic region, even as temperatures rise due to global warming, provided that fishing mortalities are reduced (Enghoff et al. 2007).

During the Little Ice Age of the late seventeenth century, coldwater marine fish (herring, flounder (*Platichthys flesus*), and eelpout (*Zoarces viviparous*)) were of major importance in the Baltic Sea fisheries and the fishing season for the major pelagic fish was substantially later in the year compared with the present, much warmer conditions (Gaumiga et al. 2007). Similarly, catches of herring and other coastal fish (for example perch (*Perca fluviatilis*) and ide (*Leuciscus idus*)) near Estonia in the mid- and late nineteenth century varied, probably owing to climatic fluctuations, when fishing effort and methods were stable (Kraikovski et al. 2008). A major hydrographic event increased the salinity of the Limfjord in 1825; the saltwater intrusion destroyed the habitat for the freshwater whitefish (*Coregonus lavaretus*), but created conditions for saltwater species such as plaice (*Pleuronectes platessa*) (B. Poulsen et al. 2007).

Overall, fishing pressure was quite low in the inner parts of the Baltic. During the late seventeenth century, removals of fish biomass from the Gulf of Riga were at least 200 times less than at the end of the twentieth century, and most fisheries concentrated on the rivers.

Migratory fish species, such as sturgeon, Atlantic salmon, brown trout, whitefish, vimba bream, smelt, eel, and lamprey were the most important commercial fish in the area, because they were abundant, had high commercial value, and were easily available. Over time, however, the main fishing areas moved downstream and to the sea. Owing to intensive fishing, populations of many migratory species, first of all sturgeon and Atlantic salmon, considerably declined and lost their commercial significance. Marine fish, especially Baltic herring, gained increased importance in the nineteenth century (Kraikovski et al. 2008).

Grand Banks, Gulf of Maine, and Scotian Shelf

Although the fisheries in the Northeast Atlantic developed later than in Northern Europe, they were no less intense in the past few centuries. The Grand Banks fishery for northern cod is a well-known example of the effects of sustained high fishing pressure ending in a sudden collapse of the stock (Myers et al. 1997).

The HMAP research focused on correcting the historical landings statistics and showed that the combined efforts of British and French fishermen on the Grand Banks off Newfoundland yielded between 204,000 and 275,000 metric tonnes of cod in the years 1769–1774 (Starkey & Haines 2001), or two to three times higher than previous estimates, and at a level that was only eclipsed by the late nineteenth century when catches were on the order of 300,000 tonnes (Cadigan & Hutchings 2001). This level proved unsustainable and catches were only half as much in the 1940s. This finding underscores that – as happened in the North Sea herring fishery – extractions using pre-industrial technology could be similar to or indeed above modern levels. When the fishery finally collapsed in 1992, landings had reached a decadal peak of 268,000 tonnes only four years earlier.

Because of the open nature of the Grand Banks fishery, the data will always be incomplete. To understand the dynamics of the fisheries fully, we need to know how many people and boats participated in a particular fishery. The focus in the later stages of HMAP has therefore been on the Gulf of Maine and Scotian Shelf fisheries closer to the American mainland that were largely conducted by local vessels through the nineteenth and twentieth centuries. Luckily, these fisheries are exceptionally well documented thanks to a bounty that required fishing captains to keep and hand in their logbooks through the period 1852–1866. Thousands of these logbooks have been digitised and analysed for content by a team at the University of New Hampshire.

The fishermen consistently removed 200,000 tonnes of live fish per year through the 1850s. For example, in 8.5 months during 1855, the hand-lining fishermen in 43 schooners from Beverly, Massachusetts, caught a little over 8,000 tonnes of cod on the Scotian Shelf, whereas in 15 months during 1999–2000 a total of just 7,200 tonnes of cod was extracted from the same waters by the entire Canadian mechanised fishing fleet and fell short of the full Total Allowable Catch by 11%, a comparison that points to a profound change in cod abundance on the Scotian Shelf over the past 150 years (Rosenberg et al. 2005).

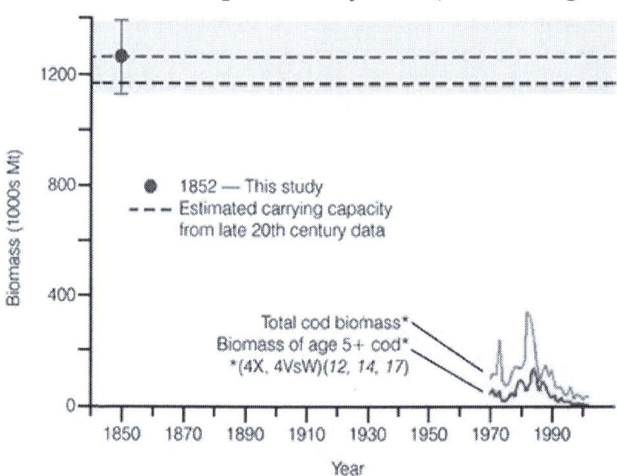

Figure: *Reduction of cod biomass, Scotian Shelf – estimated and historical. Source: Rosenberg et al. (2005). Reproduced with permission of ESA.*

Abundant as fish were, the fishermen perceived reductions in stock sizes sufficient to change to fishing grounds further at sea. By the end of the 1850s catches had declined sufficiently for many ships to undertake the longer voyage to the Gulf of St. Lawrence and the Grand Banks. Similar processes of moving from fishing ground to fishing ground in a relentless effort to earn marginal benefit are well-known for modern fisheries and well-documented for many historical fisheries, perhaps best of all for the mid-nineteenth century fishery off the Labrador coast (Myers 2001). This is a fishing strategy that is known as serial depletion and may be recognised again and again in historical records from all over the world.

Declining catches were offset by new technology. French fishermen introduced tub trawls to the Scotian Shelf fishery, and soon the Americans no longer used the traditional hand lines with two to four hooks per man but upwards of 400–500 hooks per crewman. Thus the catchment area of one boat increased immensely. Unfortunately, although catches went up in the short run, in a matter of a few years the fish stock was showing clear depletion signals, being caught at a smaller size and catch per unit effort of the fishermen declining. In the 1850s, based on the fishing effort, the adult cod biomass may be estimated to have been of the order of 1.26 million tonnes. The comparable estimate was of 50,000 tonnes in the 1990s (Rosenberg et al. 2005). The reduction in abundance is obvious and even starker than the decline of the cod and ling in the North Sea.

The American waters of the nineteenth century were incredibly rich and are today impoverished to a degree that present-day managers would not realise without historical research. It would be naive to suggest that restoration targets may simply be based on historical values. If an ecosystem regime shift has occurred, the ecosystem may never be able to rebuild to past abundance levels. However, analysis of the age structure of modern cod populations indicates that conservation measures in recent years have helped to rebuild a stock of older and better spawners, resembling the stock of the 1860s (Alexander et al. 2009).

An even more short-lived success than cod was the Atlantic halibut fishery, which became severely depleted owing to a rapidly developing taste for the halibut fins among American consumers from the 1840 to 1880s. This fishery has never regained its former strengths (Grasso 2008).

Southeast Australia

The Australian southeast shelf region was the first HMAP case study to be completed and the first case study to apply catch rate standardisation methods rigorously, single species population models, and the Ecopath ecosystem modelling approach to historical data. Compared with other HMAP case studies, the Australian southeast shelf data set is of particularly high quality. It is comparatively short in duration, beginning only in 1915 with some years missing, but it was collected in a systematic manner since the beginning of the fishery and has data for a considerable number of species. The fishery was initially set up by the government and records kept to convince private enterprise of the profitability of the industry.

What the evidence allows us to see are the effects of a trawl fishery on a pristine marine ecosystem, or as untouched by humans as was ever documented (Klaer 2005). Indigenous fishing in the Sydney region was mainly concentrated on the snapper, which lives in nearshore waters, and the indigenous fishery may have impacted the population. European settlers in Sydney added problems of pollution and disturbance. However, the southeast Australian shelf and slope marine animal populations may largely be considered to have been in a pristine state until the Australian government began fisheries experiments with a single trawler at the turn of the century. The main impact and the start of historical documentation came with the arrival of three British trawlers, purchased to begin commercial fisheries for a state company by May 1915.

The operation was privatised in 1923 and peaked with 17 steam trawlers in 1929. Danish seine vessels were brought in through the 1930s but during World War II activities almost came to a halt. Catches were resumed after the war. The fishery was primarily in shelf waters between 50 and 200 m depth. It targeted tiger flathead (*Neoplatycephalus richardsoni*), jackass morwong (*Nemadactylus macropterus*), and redfish (*Centroberyx affinis*) until the 1970s.

The unique collection of 65,000 individual haul records, vessel logbooks, and landings data were used by Neil Klaer to develop relative indices of abundance for the major commercial fish species, to estimate the biomass of those species, and to examine ecosystem changes in the southeast Australian shelf over the period since the start of commercial fishing. The results showed an overall decline in yields per haul over the history of steam trawling. Although initially the ships experienced larger catches as the men got acquainted with the new fishing grounds, the catch per hour trawled during 1937–43 was much lower than that of 1920–23. The fishing fleet moved further afield and into deeper waters as catch rates declined. In the early years, Botany Bay off Sydney yielded excellent catches of fish "very large and bursting with roe" and fishermen even talked of the "Botany Glut" from September to early December. However, by 1926 the glut failed to occur and by the 1930s the Botany Bay ground was no longer visited for commercial operations. As described for the Labrador coast, the Sydney fishermen began a mining operation that took them further up and down the coast and to deeper grounds. As the flathead was fished out, new, previously discarded fish began to be landed for the market. World War II gave a temporary reprieve for the stocks and the available biomass increased slightly, but catches quickly reduced the available biomass further.

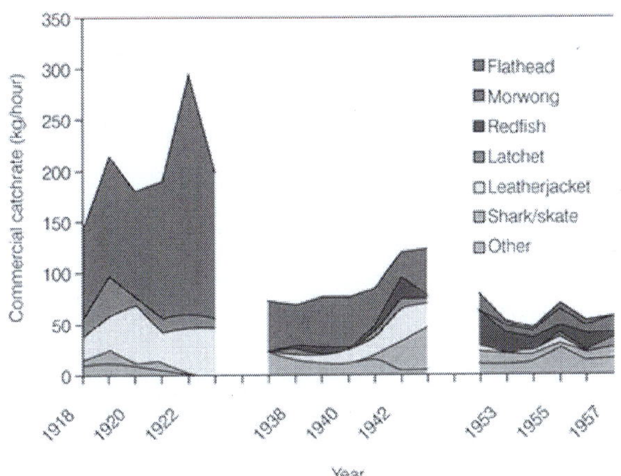

Figure: *Ecosystem effects of early trawling. Commercial catch rates by species on the southeast Australian continental shelf. Contribution per species to the total commercial catch per unit effort by year. Source: Klaer (2001).*

Despite the substantial changes in the relative biomass of the main commercial species from 1915 to 1961, there were no great changes to relative biomass at lower trophic levels (megabenthos and lower). The biomass density of all large fish (flathead, latchet, other large fish, leatherjacket, redfish, and morwong) decreased by 73% from 1915 to 1961, whereas the biomass density of invertebrates at the same time decreased by only 6% (Klaer 2004). In other words, the effect of the southeast Australian trawl fishery was a fishing down the food web, as described by Pauly et al. (1998a), resulting in fewer of the large fish, while the small fish, plankton, and crustaceans remained.

Klaer's analysis is unique both for the pristine nature of the ecosystem before documented trawling and because of the amount of data available throughout the fishery. By the time that conservation measures and restrictions on fishing effort were taken, the ecosystem had ceased to look anything like it had before, and a fishery management system that was informed only by recent data would have no knowledge of what had been lost, nor indeed of what might be if the system were allowed to rebuild by holding back on fishing effort. Until this study was made, no such information was available to managers.

Southwest Africa

Human activities were studied in the Benguela Current ecosystem (Griffiths et al. 2005). Like the southeast Australian fisheries, the main human impact occurred relatively recently, but is unfortunately less well documented. The aboriginal epoch until 1652 was characterised by low levels of mainly intertidal exploitation, whereas the pre-industrial epoch to about 1910 saw intense exploitation of few large, accessible species. The introduction of mechanised technology marked the beginning of the industrial epoch which had a huge increase in landings. Catches were stabilised in the post-industrial period after 1975, although there have been increasing impacts of non-fisheries on the system.

Total extractions in the past 200 years were calculated at more than 50 million tonnes of biomass, with annual removals above one million tonnes in the 1960s. Subsequently landings have declined by over 50%. The short, sharp impact of fisheries in the twentieth century led to severe reduction of populations of whales, seals, and pelagic and demersal fish, which are now all showing signs of recovery thanks to declining fishing pressure and implementation of new management schemes. Inshore stocks, particularly abalone, rock lobster, and inshore linefish, remain severely depressed and are exposed to intense fishery and gathering for human subsistence.

Caribbean

Far from being a pristine ecosystem, the Caribbean was intensively fished already before the arrival of the Europeans, and subsequent removals happened on a massive scale through the seventeenth and eighteenth centuries. Twentieth-century fishing pressure has continued a trajectory of fishing down the food web.

The Caribbean HMAP team has continued Jeremy Jackson's pioneering work by documenting historical distributions of large marine vertebrates. In the absence of quantitative evidence for many species, the team analysed a total of 271 descriptions from 1492 to the present, to demonstrate consistent patterns of decreased abundance and increased rarity through time. By assigning quantitative values to qualitative sightings of species, the team was able to build a comprehensive and statistically significant picture of decrease in abundance and increase in rarity in megafauna since the European settlement. The green turtle population, which had an abundance of 15.5 million to 116 million at the time of Columbus, is considered highly endangered today as only two nesting beaches with more than 100 nesting females now remain, whereas at least 23 nesting beaches have been eliminated. Similarly, 32 hawksbill turtle nesting beaches have been lost (McClenachan et al. 2006).

The decline of the monk seal followed a clear path: exploitation first reduced the range of the population, which was estimated at 0.23–0.68 million around 1500. At the beginning of the twentieth century, monk seals occupied only 30% of their former range. Hunting in the two remaining breeding areas finally killed off the monk seal from Caribbean waters by 1952.

The range of the American crocodile and the West Indian manatee was reduced in the eastern Caribbean before European settlement. Manatees were largely eliminated from the Lesser Antilles by 1700, and all but eliminated from all island sites by 1900. The decline of crocodiles was similar, except that they seem never to have been present in the Lesser Antilles. Estimates of abundance show probable declines of at least 90% for each species. The removal of large animals will have had significant consequences for food webs and the resilience of the marine system. Because hawksbill turtles consume primarily sponge matter, Bjorndal & Jackson (2003) suggest that large numbers of hawksbill turtles could have maintained high coral cover and sponge species diversity, with a concurrent increase in total benthic invertebrate diversity.

Other scientific achievements so far include an assessment of the degree of population change, ecological consequences of population change, and the historical distribution of large marine animals – including green turtles, hawksbill turtles, and monk seals – across the entire Caribbean basin and Gulf of Mexico (McClenachan et al. 2006). The team has documented and quantified changed Caribbean food webs as a whole (Bascompte et al. 2005). The Caribbean studies have brought out the potential effect of early non-mechanised fishing technologies on coral reef animals. Early declines in Jamaican coral reef fauna were a function of less than half a million people using simple gear. The research makes evident that sustainable levels of fishing in Jamaica are no more than 10–20% of current catch, or equivalent of a level of extractions already reached 100 years ago (Jackson 1997; Hardt 2008).

White and Barents Seas

The question of the impact of climate variability on fish populations is in the foreground of the work of the HMAP team working on the White and Barents Seas led by Julia Lajus. The climate effects are especially pronounced in high latitudes because many fish species occur there at the border of their distribution range. Moreover, during several centuries these effects were not masked by the human impact on ecosystems, which in the Russian north was minimal up to mid-twentieth century owing to very slow growth of the human population in the area and the late start of industrial development. Although fishing effort did increase steadily during this period, extractions were too low to influence fish populations significantly. Therefore historical data on fisheries provided a convenient research tool to trace the natural dynamics of populations, allowing reconstruction of effects of climate going back several centuries.

The team has analysed landings records from monastic and governmental sources from the seventeenth to the twentieth centuries. The records of the Solovetsky Monastery, the largest monastery in the White and Barents Sea area, which controlled sea and river fisheries of the high north, proved to be especially rich. These records are possibly unique because in many cases they contain not only the number but also the weight of caught fish: Atlantic salmon, Atlantic cod, and halibut. Atlantic salmon was one of the most valuable products of the local economy.

They were fished mostly in the lower parts of rivers, using weirs that were not changed technologically over the centuries. This makes fishing effort commensurable over time and allows comparison of historical catch data of the seventeenth and eighteenth centuries (published in D. Lajus *et al.* 2007a) with offic Analysis of historical and statistical data from four different localities around the White and Barents Seas for the seventeenth and eighteenth centuries shows a positive correlation of catches with ambient temperature (D. Lajus et al. 2005). The conclusion was drawn that before the middle of the twentieth century the population dynamics of salmon was mostly driven by natural factors (D. Lajus et al. 2008a).

Signs of climate-related dynamics were observed also for other fish, such as cod, halibut, and herring, although correlation did not approach statistical significance (D. Lajus et al. 2005, 2007b). In particular, the White Sea herring fishery, of economic importance since the eighteenth century, showed considerable short-term fluctuations of catches both because of social and natural factors and their interaction, which may confound climate effects (D. Lajus et al. 2007b). Climate effects were also pronounced on Arctic marine mammals such as white whales, Greenlandic seals, narwhals, and others, which considerably changed their distribution patterns migrating to more southern regions than usual in cold periods 1800–1809 and 1877–1903, and again in 1970–80 (D. Lajus et al. 2008b).

For marine mammals, anthropogenic pressure became a significant factor earlier than for fish. Hunting impacted the general dynamics of the population of the eastern walrus from at least the seventeenth century and may explain changes in its distribution range over several centuries. However, the walrus population was able to sustain itself as long as remote islands such as Franz Josef Land were not discovered by humans. Improvements of navigation and hunting techniques in the late nineteenth century resulted in a considerable decrease in the walrus population by the middle of the twentieth century. For fish, particularly Atlantic salmon, clear stress signals related to human activities such as overfishing and development of forestry with timber-rafting became apparent only by the end of the nineteenth century (Alekseeva & Lajus 2009).

Conducting fishing operations in such remote areas with severe climate conditions was especially difficult for humans in the pre-industrial age, causing clear interaction between natural and human factors. Fisheries productivity varied because of climate conditions, and, in particular, the price of salmon was negatively correlated with the level of catches and population abundance (J. A. Lajus et al. 2001). For herring, the long-term trend was a positive relation between catch size and human population in the area, likely reflecting an increase of fishing effort, emphasising the importance of detailed historical analysis when reconstructing long-term trends of population abundance (D. Lajus et al. 2007b).

World Whaling

Whaling was one of the most profitable extractive industries ever undertaken, and it was likely the one activity that impacted life in the oceans more than any other single pre-industrial activity. Relative to the fisheries it is extremely well documented and well researched. Yet we still do not know how many whales there used to be in the ocean and where. Whereas historical fisheries research has really only developed in the past decade, the ecological history of whaling has been pursued for management purposes for many years. Since its origin in 1946 the International Whaling Commission has had a keen interest in estimating historical population sizes based on catch records in order to identify a conservation target for the rebuilding of whale populations. The approach taken by the HMAP team is to estimate historical abundance using population models based on the evidence of historical logbooks and catch records, and using present-day abundance estimates.

A global overview of the history of whaling identified 120 whaling operations grouped into 14 methodology-defined eras (Reeves & Smith 2006). Maps of the spatial and temporal extent of whaling in the nineteenth century allow resource managers to identify areas where populations have and have not recovered to their pre-whaling distribution, and to identify formerly occupied areas where whales are now essentially absent. Where recovery has been less than complete, human activities may need to be better managed to allow further recovery. Spatial distribution should become a more important element in assessing population recovery, in addition to the more usual measure based on current population size as a fraction of historical, or pre-whaling, population size.

The catch history of North Atlantic humpback (*Megaptera novaeangliae*) whaling was estimated by the HMAP team (Smith & Reeves 2006), and used in a stock assessment sponsored by the International Whaling Commission to estimate that current abundance is 37% to 70% of the historical abundance, which itself was 22,000–26,000. A previously unknown humpback whale feeding ground was identified based on nineteenth century whaling logbooks (Reeves et al. 2004), and the logbook data were used to help direct the Census Patterns and Processes of the Ecosystems of the Northern mid Atlantic (MAR-ECO) project. In the mid-nineteenth century some humpback whales migrating from breeding to feeding areas remained at mid summer in oceanic habitats near the mid Atlantic Ridge. Today humpbacks have only been known in summer months on coastal feeding grounds around the North Atlantic. Similarly, textbook assumptions on the distribution and abundance of North Pacific right whales (*Eubalaena japonica*) have been corrected (Josephson et al. 2008). The causes of failure of the North Pacific right whale to recover both numerically and spatially after the severe depletion of the 1840s continue to be a mystery.

Working as part of the HMAP New Zealand project, the team described the historical distribution and landings of southern right whales (*Eubalaena australis*) through analysis of over 150 whaling logbooks and other landings records. With 95% statistical confidence, population modelling shows that southern right whales numbered between 22,000 and 32,000 in the early 1800s, declining rapidly once whaling began. By 1925, perhaps as few as 25 reproductive females survived. Today the population has recovered to some 1,000 animals around sub-Antarctic islands south of New Zealand (Jackson et al. 2009).

Because of the strong need to establish past population sizes to inform conservation policy, the HMAP team is working closely with scientists who have proposed another possible modelling approach to working on historical landings data. This is the Whales Before Whaling project headed by Steven Palumbi at Stanford University. Palumbi's project aims to measure the amount of genetic diversity of current populations and use knowledge of DNA mutation rates to estimate how many individuals a population must sustain over time to accumulate the measured diversity. Based on this method, Palumbi estimates that the pre-contact population size of the eastern Pacific gray whales (*Eschrichtius robustus*) was three to five times larger than the population size calculated by historical data. The HMAP team has therefore scrutinised the available landings and total removals and, in cooperation with the US National Marine Fisheries Service and the International Whaling Commission's Scientific Committee, is working to address this apparent inconsistency between historical whale removals, apparent population increases measured over the latter half of the twentieth century, and the genetic variability model (T. Smith, personal communication).

We now know more about the human drivers behind the whaling operations. In particular the project focused on how the enormously profitable so-called Yankee whaling changed from 1780 to 1924. The team has documented the nature of the changes in vessels, rigging, destinations, and catches over the lifespan of this fishery. They suggest that questions of the effect of whaling on the whale populations must be asked at regional rather than global levels, and that indeed regional depletion, even extirpation, was a frequent occurrence.

For example, contrary to Whitehead (2002), they show strong depletion of sperm whale abundance in the Pacific and raise the question why such depletion apparently did not occur in the Atlantic (Smith et al. in press). They also suggest that global economic nalyses that do not account for these regional changes greatly oversimplify the dynamics of this fishery and are misleading about the causes of its decline.

Megamollusks

Shellfish have been heavily collected and used for meat and ornaments through history. Although some shells will be traded, most will be discarded. Shell middens have been known since the nineteenth century as excellent archaeological sources of information on coastal-dwelling peoples. Through the nineteenth and twentieth centuries, the oyster reefs of the US Atlantic and Pacific coasts were severely impacted by fishing (Kirby 2004) as were North European oyster banks (Holm 2005).

Thanks to the initiative of Andrzej Antczak of Venezuela, we now have a global series of studies of human megamollusk interactions. Generally, mollusk populations are quite exposed to human impact as they may be collected close to the shoreline. The southwest African HMAP project showed that human gathering of inshore shellfish may reach a level where it threatens certain inshore species. In Papua New Guinea the exploitation of the giant clam (family Tridacnidae), which seems to have been at sustainable levels through a long period of history, has in recent decades necessitated a ban on collecting for export (Kinch 2008).

Similarly, although ecological impacts such as declining size may be detected for the pre-Hispanic exploitation of queen conch (*Strombus gigas*) beds off Venezuela, the exploitation was much less harmful to the mollusk population than the short-term modern fishery between 1950 and the 1980s (Antczak et al. 2008). However, few human populations have been so dependent on mollusks for food to cause local or species extinctions (Bailey & Milner 2008).

The study of megamollusks is particularly rewarding for our understanding of human values and trade. The queen conch was heavily targeted between about 1100 and 1500 at the offshore islands of Los Roques, Venezuela, and both the meat and shells were brought to the mainland for consumption and redistribution. Ceremonial activity on the islands and the use of the queen conch as a symbol on the mainland indicate that the mollusk had achieved a central importance to north-central pre-Hispanic peoples in Amerindian Venezuela (Antczak and Antczak 2008).

Emerging Projects

Two HMAP projects have not yet arrived at publication stage because they were only begun fairly recently: the New Zealand and the Southeast Asia projects. The Maori were experienced sailors and hunters, and on their arrival to New Zealand the Europeans encountered nothing like a pristine ecosystem. The Taking Stock project is therefore confronted with understanding fully the impact of pre-European, pre-industrial technologies on what was, until the arrival of the Maoris around 1300, a pristine marine and terrestrial ecosystem. The project will conclude by the end of 2010, but it is clear that the distributions and population sizes of seabirds, fur seals, and sea-lions were considerably impacted relative to pre-Maori conditions already by 1800.

Fur seals, for instance, had been extirpated from North Island and only colonies at the southern tip of South Island awaited the arrival of European hunters to be rendered extinct. The Southeast Asia project covers a vast area and focuses on indigenous and American historical whaling in the Philippines, Taiwanese offshore tuna fishery, and shark fishing in Indonesia. All HMAP Asia research projects are now at an advanced stage, and a monograph (representing the main output of the project) is being prepared for publication by the end of 2010.

Conclusions

What is the big picture emerging from these regional and species projects? What is the scale of change between now at the completion of the Census and, say, 100 years, or between now and the origin of large-scale pre-industrial fisheries? When were the decisive moments? What were the main drivers? These are questions that we are grappling with now as the Census is coming to an end. Already we know some of the answers but many more will emerge as we have an overview of the vast amount of information that has been uncovered.

- The HMAP project has resolved the problem of the baseline. We now know that everywhere we look there is potential to know much more about the past and that we need to inform ourselves of the past both to enrich our understanding of the present and to inform our future preferences and decisions. The HMAP project is the beginning of the historical discovery of ocean and human interaction. Even after 10 years we have far from exhausted the archives and archaeology of the sea. We have made significant discoveries both of the importance of the sea to human life and of the impact of humans on the sea. Historical baselines should be an important element of future conservation plans. In some ecosystems, stocks will rebuild if given a chance. In other systems, regime shifts may have forever changed the food web so that past abundances of top predators will have a slim chance of rebuilding. Yet, environmental history has a very real role to play for future ocean policy by preserving the memory of what once lived in our seas. New management policies can be developed to promote recovery and prevent further declines of species and ecosystems.

- The distribution and abundance of marine animal populations change dramatically over time. The effects of climate variability during the Little Ice Age on marine mammals as well as fish stocks are clearly documented by the White and Barents Seas project and the Baltic Sea project, whereas the North Sea documents the effects of the past 20 years of warmer surface water for the introduction of southern species. Historical data will inform us of past patterns of distribution of species such as demonstrated for North Atlantic humpback whales and North Pacific right whales, and indeed for the southern North Sea haddock.
- We now know that major extractions occurred more than 2,000 years ago in the Mediterranean and Black Seas, we know the basic outline of the origins of commercial fisheries in Northern Europe, and we have a good sense of developments in many regions around the globe during the past 500 years ranging from the Caribbean to the White Sea, from southeast Australia to South Africa. Pre-industrial technologies were sufficient to put marine animal populations under severe stress, and indeed by the late nineteenth century extractions in Europe, North America, and the Caribbean had reached levels that would be equivalent to today's Total Allowable Catches. The effects of large-scale removals in the seventeenth century North Sea herring fishery and the eighteenth century Grand Banks cod fishery may have been significant.
- Regime shifts may have occurred as a result of some pre-industrial fisheries such as the Caribbean, whereas the effects of industrial gear were striking in the southeast Australian case when a pristine ecosystem changed dramatically after 30 years of trawling. Collapses of stocks and serial depletion are widespread phenomena, even before the industrial era, but in most cases populations have been able to rebuild.
- Overall it seems that the removals of large marine animals have reduced abundance by an order of magnitude; a recent review concluded that 256 exploited populations declined 89% from historical abundance levels on average (range: 11–100%)

(Lotze & Worm 2009). The detailed historical evidence for cod, ling, and bluefin tuna corroborate this general picture. Smaller animals have been less impacted and indeed may have replenished as larger predators have been removed.

- Human impacts on coastal environments have been similar across the globe, even in quite different ecosystems (Lotze et al. 2006). Although few exploitable marine species have gone extinct, there is concern that entire marine ecosystems have been depleted beyond recovery. Major impact on sensitive ecosystems such as the Wadden Sea may have happened before 1900, and there is therefore a need for deep historical assessment of ecosystem change.

We know now that we can push back the chronological limits of our knowledge.

More importantly perhaps, we now have the basis from which to start raising new questions: what more can we know about the drivers of change from the human perspective, can we extrapolate from the local or regional to the global, what about the continents or large countries that did not have an HMAP team such as much of South America and Africa, what about India and China? Can we unlock the sources to some of the large industrial fisheries in the deep seas that have become such important fisheries areas in recent decades for already endangered species such as orange roughy and Patagonian toothfish?

All these are challenging but certainly not impossible questions. They are questions that will not only be raised but answered in coming years as research continues beyond the HMAP project. Data rescue and digitization will provide vastly increased libraries of the kind we already know and that have served us well. We are beginning to understand the main drivers of change from the human perspective such as changing patterns of consumption, technology, price differentials, politics, and cultural preferences. We now know enough to begin to understand the importance of marine products for human consumption, and we have a much better basis from which to assess the main drivers of human marine exploitation. In the academic realm the historical turn of marine ecology is now a given, and the ecological challenge to traditional historical models cannot be neglected. In the future, new techniques and methodologies may be used to move what may now seem the unknowable into the realm of the knowable. If knowing the basics of marine ecological history seemed impossible some 10–15 years ago, we stand a good chance that in the next 10–15 years there will be several major breakthroughs. Scientific advances in fields such as genetics and stable isotope analysis have already impacted what we know and much more is to come. Advanced computer animations and geographic information systems (GIS) will be fully used to show changes in abundances, distributions over time, and how they could look in the future (under recovery situations), and we shall see new quantitative approaches for modelling changes in biodiversity, species' abundance, and distribution.

Perhaps new methodologies will enable us to lift the veil on what the pristine sea looked like before human contact. So far nearly all the information accessible to us relates to early human records of contact. Certainly we shall know much more about the implications of the ice ages for "trapped" species. Advances of molecular biology and ocean biogeography will tell us of the separation of species and subsequent development. Sediment cores will be unlocked as a library of past DNA of what used to be swimming in the water column above. All of this will underline what was one of the first steps of the HMAP project, the need to train the next generation of researchers in interdisciplinary skills.

Surveying Nearshore Biodiversity

The nearshore region is defined here as the area from the high intertidal down to 20 m water depth, which is the focus of the Census ofMarine Life Natural Geography in Shore Areas (NaGISA) project. The overarching goal of NaGISA is to produce nearshore biodiversity baselines with global distribution from which new scientific questions and hypothesis testing can arise, long-term monitoring can be designed, and management plans can be implemented. One of NaGISA's goals is to create accurate biodiversity estimates by producing species lists for nearshore sites around the world. Previous overallmarine biodiversity estimates, which include the nearshore, range from 178,000 to more than 10 million species (Sala & Knowlton 2006).

To narrow this large range and obtain specific assessments for the nearshore, more species lists from more nearshore regions of the world are needed such as those produced during the NaGISA project. One example of the use of NaGISA baseline data is to examine latitudinal trends in biodiversity. Thus far, there have been few truly global nearshore biodiversity comparisons attempted because of the lack of comparable data (e.g. Witman et al. 2004; Kerswell 2006). NaGISA contributes to our ability to make latitudinal and other spatial comparisons by establishing a standardised samplingprotocol ensuring comparability of datasets and by greatly increasing the data coverage over a large latitudinal and longitudinal range. NaGISA also has initiated a growing network of scientists that will continue to accumulate data in the years to come. This project and its goals are particularly timely because of the changes in nearshore biodiversity that are resulting from increasing anthropogenic impacts and the changing climate.

NaGISA is a Japanese word that translates into the "area where the sea meets the land". Specifically, the goal of NaGISA is to assess nearshore biodiversity in rocky macroalgal and soft-bottom seagrass areas from the high intertidal to a water depth of 20 m. Within NaGISA these nearshore habitat types were chosen for two reasons. First, these habitats are known to have high biodiversity because of the three-dimensional structure provided by the macrophytes. Even in nearshore areas where soft sediments dominate, small macrophyte oases have a higher biodiversity than the surrounding soft sediments (Dunton & Schonberg 2000). Second, these habitats are fairly globally distributed, in contrast to other habitats like nearshore coral reefs that are typically restricted to warmer waters.

One of NaGISA's largest legacies is the development of a standardised sampling protocol for nearshore rocky macroalgal and seagrass habitats (Rigby et al. 2007). This protocol ensures comparability among all NaGISA data to make an evaluation of large-scale to global nearshore biodiversity patterns possible.

In addition to data comparability, a major hurdle for many nearshore biodiversity surveys is a lack of taxonomic information for many groups beyond conspicuous macrofauna and flora, especially for many smaller organisms that make up much of the existing biodiversity. NaGISA's network of scientists includes local taxonomists as well as taxonomic training to ensure accurate and reliable identifications for all major taxonomic groups. However, given the comprehensive coverage resulting from NaGISA collections, a lack of taxonomic expertise still exists for many of the smaller and less charismatic organisms and in many regions of the world.

For organisational purposes, NaGISA divided the world's shorelines into eight regions: Western Pacific, Eastern Pacific, South American Seas, Caribbean Seas, Indian Ocean, Atlantic Ocean, European Seas, and Polar Seas. As of May 2010, the NaGISA project has sampled 253 sites, of which 179 were macroalgal sites, 71 were seagrass sites, and one each was a rhodolith site, a sandy beach, and a mudflat. NaGISA also organises the world's coastline into 20 degree bins and has data coverage (at least one sampling site per bin) in about 45% of these nearshore bins so far. Also, of the 253 sites, 64 sampled so far have been sampled more than once and many are on their way to becoming long-term monitoring sites.

This initial census (2000–2010) provided a baseline dataset for long-term monitoring and the information needed to answer fundamental ecological questions about spatial patterns in nearshore biodiversity. Building on this growing baseline, NaGISA data will eventually help identify the drivers that structure these nearshore communities on local, regional to global scales. Apart from its scientific value, the strength of NaGISA is that it involves local interests and stakeholders, from local community groups to elementary, high school, and university students. This allows stakeholders to become vested in the nearshore and build an on the ground force that uses NaGISA data to solve local management problems. NaGISA data are part of the OBIS database and are thus publicly available. As of May 2010, NaGISA contributed over 47,700 records towards OBIS distributional maps with a total of over 3,100 taxa.

Table : *Sites sampled by NaGISA by region and habitat type. A total of 253 sites have been sampled during NaGISA activities in macroalgal and seagrass habitats. Other habitat types include rhodolith beds, mudflats, and sandy beaches. Data as of May 25, 2010*

	Sampling effort		***Site habitat types***		
Region	***Point data sites (single sampling)***	***Monitoring sites (multiple samplings)***	***Macroalgal***	***Seagrass***	***Other***
Atlantic Ocean (AO)	8	5	5	7	1
Caribbean Sea (CS)	59	22	45	35	1
Eastern Pacific (EPAC)	45	13	51	6	1
European Seas (ES)	5	4	8	1	0
Indian Ocean (IO)	32	7	34	5	0
Polar Seas (PS)	8	11	19	0	0
South American Seas (SAS)	4	2	4	2	0
Western Pacific (WPAC)	28	0	13	15	0
Total	189	64	179	71	3

The Status of Regional Nearshore Biodiversity Knowledge

The nearshore region is highly accessible and, as such, has historically received much taxonomic and ecological attention from scientists and naturalists. As with other ocean biomes, taxonomic and biodiversity knowledge differs depending on geographic region and taxonomic group. Even with this varying knowledge base, nearshore field guides and scientific publications exist for most regions of the world. It is therefore surprising that before NaGISA, very few regional estimates for nearshore biodiversity existed and information regarding biodiversity patterns on the regional scale was scarce. The following is a brief highlight that describes the status of nearshore biodiversity knowledge in each of the eight NaGISA regions.

Eastern Pacific (EPAC)

NaGISA sites sampled in the Eastern Pacific region span from approximately 61° N (south–central Alaska) to 24° N (Baja Mexico). Fifty-eight sites have been established in various locations along the coasts of the United States (Alaska and California), Canada (British Columbia), and Mexico (Baja). Of these sites, 13 have been sampled more than once and are becoming established monitoring sites. Some sites in Alaska were established with the assistance of local native communities, and some sites in both Alaska and California are being maintained with the assistance of various high school and university classes.

Although much research has been done in this relatively well-known region, there are no estimates for overall nearshore biodiversity. Nonetheless, some latitudinal descriptions of this region do exist. Early work demonstrated that benthic processes, such as competition and predation, caused a north–south gradient of decreasing recruitment of intertidal sessile invertebrates from Oregon to California (Connolly & Roughgarden 1998). Along the Pacific coast of North America biogeographical and oceanographic discontinuities separate rocky intertidal communities into 13 distinct spatial groups (Blanchette et al. 2008). In general, they found strong correlations between species similarity and both geographical position and sea surface temperature. Supporting this view is the observed latitudinal gradient in the recruitment of intertidal invertebrates for this region (Connolly et al. 2001). Interestingly, in this same region, Schoch et al. (2006) suggested that wave run-up was the most significant physical parameter that affected community structure. NaGISA has added much knowledge to this region by starting the first extensive nearshore monitoring in

Alaska and by adding to existing datasets, which will allow for a more complete longitudinal comparison along the Northwestern American coast

Western Pacific (WPAC)

NaGISA sites in the Western Pacific region span from approximately 43° N (Eastern Hokkaido, Japan) to 8° S (Indonesia). Twenty-eight sites have been established in various locations in Japan, Vietnam, Philippines, Thailand, Malaysia, and Indonesia. Although so far none of these sites has been sampled more than once, current WPAC efforts are trying to establish several monitoring sites.

Although much research has been done in this region, particularly in Japan, there are no nearshore biodiversity estimates. Nonetheless, some latitudinal descriptions do exist along some major ocean current regimes. Along the northern Japanese coast, the subarctic, southerly flowing Oyashio current is characterised by high biomass, large individuals, and low biodiversity. In contrast, the warm, northerly flowing Kuroshio current along the southern Japanese coast is characterised by high biodiversity but low biomass (Nishimura 1974). The high biodiversity in the Kuroshio region occurs because this current transports species living in the high diversity Coral Triangle around the Philippines, Indonesia, and Malaysia to the northern subtropical and temperate regions of the western Pacific. The high biodiversity in the south Asian coastal area has sparked much research, including important taxonomic work. NaGISA has contributed to some of these publications, such as field guides on echinoderms (Yasin et al. 2008), hermit crabs (Rahayu & Wahyudi 2008), and seagrasses (Susetiono 2007).

European Seas (ES)

NaGISA sites sampled within the European Seas region span from approximately 55° N (Poland) to 35° N (Crete). Sampling sites have been established in the North Sea, the Baltic Sea, the East Atlantic Ocean, the Northwest Mediterranean, the Northern and Southern Adriatic Sea, and the Aegean Sea, with collaborators from Italy, the United Kingdom, Portugal, Greece, and Poland. A total of nine sites have been sampled, four of which have been sampled more than once.

Although the biodiversity of individual regions within the European Seas has been the focus of intense research (Frid et al. 2003), an exhaustive analysis of biodiversity estimates, patterns, and trends is lacking. One pattern that has been noted is the replacement of large canopy algae that dominate at higher latitudes with seagrasses that become dominant in the Mediterranean, where relict kelp populations persist only in the Strait of Messina and in the Sicily Channel (Lüning 1990). NaGISA information in the ES is allowing researchers to explore nearshore processes more thoroughly than before. For example, NaGISA data have helped to show that rare species may become more abundant when the environment is variable (Benedetti-Cecchi et al. 2008)

Indian Ocean (IO)

The Indian Ocean NaGISA sites range latitudinally from 28° N (Egypt) to 34° S (South Africa) and are found in Kenya, Tanzania, Mozambique, India, Egypt, and South Africa. Of the 39 sites that have been sampled, seven have been sampled more than once and are on their way to becoming monitoring sites. Two of the sites in Tanzania were established and are being monitored with the assistance of high school students, both local and from the United States.

As a result of several landmark expeditions and later research, taxonomic knowledge of the Indian Ocean region has been expanding. However, although biodiversity estimates do exist for certain groups in particular areas, latitudinal biodiversity descriptions for this region are lacking. The southern region of the African continent is particularly high in coastal biodiversity, with estimates of over 12,000 species from southern Mozambique in the Indian Ocean to northern Namibia in the east Atlantic, representing 6% of all coastal marine species known worldwide (Branch et al. 1994; Gibbons et al. 1999; Adnan Awad et al. 2002; Griffiths 2005). Other coastal regions of the IO are largely unknown, such as the island marine fauna in India, which have been estimated to be approximately 75% unknown (Venkataraman & Wafar 2005). In the IO region, NaGISA efforts are focusing to contribute specifically to areas of currently little existing information such as India.

Atlantic Ocean (AO)

The Atlantic Ocean region was sampled at 13 sites ranging from approximately 47° N (Canada) to 13° N (Senegal). These sites have been located along the coasts of Canada, the United States (Maine to Connecticut), and Senegal. Sites in Canada and the United States have largely involved elementary, high school, and university students for their sampling. Of the AO sites, five have been sampled multiple times and are considered monitoring sites. In 2010, at least 12 additional sites will be established and monitored in collaboration with summer science camps from Connecticut to Maine in the Unites States.

In the AO region, it is generally recognised that biodiversity increases with decreasing latitude when comparing boreal with tropical regions (Udvardy 1969). Various environmental factors, such as local habitat heterogeneity can complicate this trend at the local scale. For example, NaGISA sampling has helped to show that Cobscook Bay at the US/Canada border, contrary to the general trend, has substantially higher macroinvertebrate species diversity than areas further south (Trott 2009). In addition, there are distinct biogeographic regions in the Northwest Atlantic, including the Polar, Acadian, Vrginian, and Carolinian Provinces, with distinct regional diversity patterns (Pollock 1998).

South American Seas (SAS)

The South American Seas sites extend from a latitude of 2° S (Ecuador) to 42° S (Argentina) and include the countries of Argentina, Ecuador, and Brazil. A total of six sites have been sampled, with both Argentinean sites being sampled twice. All sites in the SAS region were sampled with the assistance of local university students.

Although much local knowledge exists within various countries in this region, good nearshore biodiversity estimates and discussions of latitudinal trends are scarce. In Brazil, 540 taxa were described associated with seagrass beds, mostly polychaetes, fish, amphipods, decapods, mollusks, foraminiferans, macroalgae, and diatoms (Couto et al. 2003). Other areas, such as the fjords in southern Chile, have received little attention so far, and recently explorations have discovered 50 new species associated with them (Haussermann & Forsterra 2009). In Chile, several marine invertebrate taxa were found to decrease in biodiversity with increasing latitude between 18° and 40–45° S, and then increase further south, probably because of the presence of sub-Antarctic fauna (Gallardo 1987; Clarke & Crame 1997; Fernandez et al. 2000). NaGISA is contributing to the overall biodiversity effort in the SAS region by attempting to establish well-distributed NaGISA sites that will greatly enhance communication among countries so that larger-scale comparisons can be made.

Caribbean Sea (CS)

The Caribbean Sea sites span from approximately 10° N (Venezuela) to 30° N (Florida). Although latitudinally this is the shortest NaGISA region, it has an impressive total of 81 sites from the countries of Cuba, Trinidad and Tobago, Venezuela, Colombia, and the United States (Florida). Of the 81 sites, 22 have been sampled more than once. Many of the sites in Venezuela have involved university students in their sampling, and the Florida site was initiated by a high school group, which has also gone on to help other high school groups with NaGISA sampling around the world, including Greece, Zanzibar, and Egypt.

It should be noted that for the Caribbean Seas, NaGISA is the first attempt to establish a monitoring program that does not target coral systems. This is particularly important for this region because the massive changes that have occurred in coral reefs over the past several decades (Gardner et al. 2003), including an 80% drop in live coral cover in 25 years (Wilkinson 2004), have prompted an increase in hard substrate availability, which in turn might result in a phase shift from coral-dominated communities to hard-bottom macroalgal communities.

With the exception of general field guides and some specific scientific publications, no nearshore biodiversity estimates or biodiversity trends are known to exist. However, NaGISA is contributing to this knowledge, by producing the first longitudinal comparison in the CS region, which has shown that diversity decreases from west to east.

Polar Seas (PS)

The Polar Seas region includes both the Arctic and the Antarctic. There are 13 Arctic NaGISA sites that were sampled around 70° N, off the United States coast of Alaska. Eight of these sites have been sampled multiple times and are monitoring sites. In the Antarctic, six sites have been sampled at 62° S and 78° S. Five of these sites were off of the United States McMurdo Station and one was off of the Uruguayan Artigas Research Base at the Antarctic Peninsula. Of the sites around McMurdo Station, three have been sampled more han once.

Biodiversity estimates are scarce for both polar regions for mosttaxa. However, for macroalgae it is estimated that there are as many as 120 macroalgal species in the Antarctic (Wiencke & Clayton 2002) and slightly more in the Arctic (Wilce 1997) but with a much higher percentage of endemic species in the Antarctic. The polar regions also have little information available regarding latitudinal trends. Typically, Arctic nearshore systems are thought to be less diverse than northern temperate systems. In the Arctic nearshore, it seems that higher diversity is typically found at more southern locations compared with northern locations. In the Antarctic, the Peninsula, which spans approximately six degrees of latitude from 62° to 68° S, shows a latitudinal macroalgal decline (Moe & DeLaca 1976). Extending this gradient further south to the Ross Sea (77° S), the southernmost location of open water, only two species of fleshy macroalgae occur (Miller & Pearse 1991). This latitudinal decline is mainly driven by reduced light availability with increasing latitude due to strong seasonality, low solar angle, and extended periods of ice cover.

Biodiversity Gradients

Latitudinal gradients of increasing species diversity from the poles to the tropics have often been touted as a fundamental concept in terrestrial ecology (Willig et al. 2003). Many mechanisms have been proposed to explain this latitudinal gradient, but changes in temperature have been targeted as the most plausible factor in terrestrial systems. The variation in ocean temperatures over the same distance, however, is significantly smaller and the overall importance of temperature versus other physical factors has only begun to be discussed (Blanchette et al. 2008). Other mechanisms driving latitudinal trends of rocky nearshore biodiversity are primarily large-scale oceanographic conditions and local biological interactions, which can include nutrient content and, thus, primary productivity, local assemblages of herbivores and predators, the prevalence of larval stages with differing dispersal ranges, speciation rates, and so forth (Connolly & Roughgarden 1998; Roy et al. 2000; Broitman et al. 2001; Connolly et al. 2001; Rivadeneira et al. 2002; Okuda et al. 2004; Kelly & Eernisse 2007).

Debate still surrounds the existence of nearshore latitudinal biodiversity trends, especially on the global scale. The reason for this is the lack of studies actually completed at the global scale. It is time intensive and costly to sample sites globally and literature reviews are difficult to compare owing to the various biases associated with using different sampling protocols. Even with these constraints, there are two excellent examples of global studies. In one study, field sampling found that shallow subtidal boulder communities tended to havehigher species numbers at equatorial sites compared with sites closer to the poles (Witman et al. 2004). In contrast, a study based on a literature search of nearshore algal genera found that more biodiversity hot spots occurred in temperate regions compared with tropical or polar (Kerswell 2006). Although both studies are ground-breaking as they were the first to attempt global comparisons, it should be noted that they are limited in that one was completed on a specific habitat (subtidal rock walls in 12 biogeographic regions, totalling 49 local sites) and the other focused on one taxonomic group (macroalgae). NaGISA is assisting to broaden the knowledge of global biodiversity by increasing the number and distribution of sites, increasing the range of habitats (including intertidal and subtidal rocky shores and seagrass beds), and increasing the number of taxa examined. Based on NaGISA's main target taxa, global latitudinal comparisons will be possible for macroalgae, seagrasses, mollusks, echinoderms, polychaetes, and decapods, in addition to comparisons of overall community composition in rocky shores and seagrass systems.

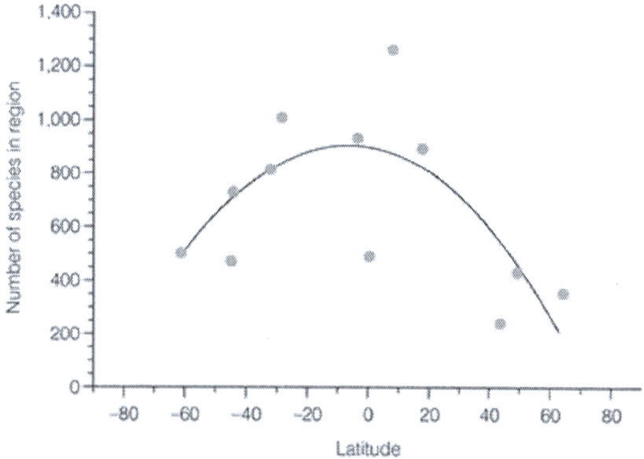

Figure: *Regional species richness as a function of latitude. Reproduced with permission from Witman et al. (2004). Copyright 2004 National Academy of Sciences, USA.*

Biogeographic Breaks

We cannot discuss biodiversity gradients without mentioning biogeographic breaks. iogeographic breaks are important because biodiversity gradients do not always change continuously but sometimes are abrupt owing to these breaks. Breaks can be driven by the dynamic interaction of two or more disinct water masses. This creates active transition zones where species mingle across their respective boundaries, for example, the biogeographic provinces associated with cold- and warm-water masses. These transition zones include species pools from both systems, often resulting in a high level of biodiversity at the breaks.

Biogeographic breaks are worldwide. For example, in the east Pacific, a well-studied biogeographic break is Point Conception in California. Offshore of Point Conception, the continental shelf is broad and the south-flowing California Current is deflected offshore (Brink & Muench 1986; Browne 1994). Point Conception is a "transition zone" between the warm Californian Province and the cooler water regime of the Oregonian Province, resulting in different fish, invertebrate, and algal communities on either side of this break (Horn & Allen 1978, Murray & Littler 1981; Murray & Bray 1993). Similarly, in the eastern Atlantic along the western African coast, the coastal waters of Mauritania and Senegal and adjacent areas form a transition zone between a more temperate northern zone and a warmer tropical zone farther south.

Despite variations in local conditions, biodiversity patterns of fishes, invertebrates, and particularly macroalgae reflect this change within a relatively narrow 400–500 km band (Lawson & John 1987). For Eastern South African macroalgae, a biogeographic break occurs at St. Lucia, 135 km south of the Mozambique border. Here, there is a transition from a tropical Indian Ocean flora to a temperate South African flora. As another example, a biogeographic break is found in the Gulf of Maine at Penobscot Bay, Maine, where the Maine coastal current splits to flow southwest from eastern Maine. One of the resulting branches travels east and the other continues in a southwestern direction. The communities above and below this break are statistically distinct, but not within either of the two regions. The already mentioned boundary of the subtropical, warm Kuroshio current and the subpolar, cold Oyashio current forms an important biogeographic break along the eastern coast of Japan, influencing patterns of diversity and biomass. There are other biogeographic breaks around the world; these are just a few to highlight their importance to biodiversity.

Some biogeographic breaks are still under investigation and highlight the need for more biodiversity studies. For example, in the Aleutian Archipelago in Alaska, a biogeographic break may exist that drives the presence of the canopy-forming kelp from only *Eualaria fistulosa* to the west to primarily *Nereocystis luetkeana* to the east (Miller & Estes 1989). However, more oceanographic and biological data are needed to identify the exact location and drivers of this possible break (Ladd et al. 2005). NaGISA is assisting in this discussion by establishing sites along the Aleutian Archipelago.

Nearshore Biodiversity Hot Spots

A biodiversity hot spot is a biogeographic location that contains an unusually high number of species. Hot spots may occur along a coastline where habitats are homogeneous but for some reason a particular location has high biodiversity. A hot spot also may occur at a site where habitat type is different than the surrounding environment, as commonly seen in deeper waters at seamounts surrounded by soft sediment. There are many reasons why species diversity may be higher in certain locations and these reasons are often site specific. Reasons may include change in substrate, water mass, topography, nutrient intrusions, or geologic history.

An example of a NaGISA site that is a hot spot because of a substrate change is in the Arctic Beaufort Sea (in the PS region). Here, the typically soft-bottom seafloor contains a low-diversity fauna, with only about 30 infaunal species, mainly polychaetes and amphipods (Feder & Schamel 1976; Carey & Ruff 1977; Carey et al. 1984). In this region, local biodiversity hot spots occur where boulders provide colonisable hard substrate for macroalgae and sessile epibenthic macrofauna, which attract other organisms including more than 150 species of macroalgae, invertebrates, and fish (Dunton et al. 1982).

Hot spots also can be created by oceanographic conditions, such as in the Gulf of Maine (Buzeta et al. 2003; Trott & Larsen 2003). The NaGISA site in Cobscook Bay has the highest species richness of macro-invertebrates of any bay similar in size and habitat characteristics in the Gulf of Maine, with approximately 800 known species representing all major phyla (Trott 2004). The high biodiversity of Cobscook Bay appears to result from wave exposure and the extraordinary tides this system experiences (Campbell 2004). Additional hot spots were also identified in the Bay of Fundy where NaGISA assisted the Department of Fisheries and Oceans Canada in an effort to determine Ecologically and Biologically Significant Areas (EBSA), which resulted in the identification of five EBSA's in the Quoddy Region (Buzeta & Singh 2008).

Closing Information Gaps

The field of taxonomy, traditionally based primarily on morphology, has expanded in recent years to include molecular information (Blaxter 2003; Hebert et al. 2003). This has not only enhanced our understanding of evolutionary relationships but also our knowledge of biodiversity and species distributions ranging from algae to fishes (Saunders 2005, 2008; Blum et al. 2008; Pfeiler et al. 2008; Thacker 2009). Nonetheless, our ability to identify organisms in some areas and for some taxa is still limited, leaving gaps in taxonomic knowledge as well as for particular regions of the world's coasts. Many developing countries and remote regions lack financial support, technology, and taxonomic information for their fauna and flora. This is particularly true for smaller, less charismatic organisms of no economic importance. Access to these remote regions is difficult due to logistical and financial constraints, leaving gaps in data coverage. Each of the eight NaGISA regions contains areas that have not been sufficiently explored. The western coast of Alaska in the Eastern Pacific region, the western African coast in the Atlantic region, all of the Arctic coastline except where research stations allow access, the eastern Antarctic coast, and remote islands in the Indian Ocean are just a few examples.

NaGISA's Major Findings

Although the nearshore region is probably among the most-studied parts of the ocean because of its accessibility and obvious interest to humans as a resource, the lack of information on biodiversity and its large-scale and long-term patterns in more than a handful of locations is particularly surprising. Also surprising is the lack of integrated information so that regional and global trends and patterns can be discussed. NaGISA is the first project to undertake the ambitious step to create such large-scale baselines with the establishment of standardised protocols and a growing global network of nearshore researchers. With over 250 sites located around the world, and still growing, and 28 different countries involved, NaGISA is the largest-ever attempt to address truly global-scale biodiversity issues.

The central idea of the NaGISA standardised sampling protocol is a fully nested design. Replicate samples along various tidal heights are collected at each site, and multiple sites are sampled within regions of specific latitude and longitude. This hierarchical design of the protocol with replicate samples within a site, which is then nested within latitude or longitude, allows a statistically appropriate and powerful method to analyse biodiversity patterns across several spatial and temporal scales (Benedetti-Cecchi 2007). Not only can biodiversity patterns be analysed on local, regional, and up to global scales, but it can also be determined at which of these scales most variability occurs.

NaGISA's protocols include various independent sampling levels, from cover estimates to actual collections and detailed taxonomic identification of all organisms. This design allows flexibility in sampling effort, so where the full sampling effort is not possible due to logistical or financial constraints, parts of the protocols can be used to create important local information that can be compared with large-scale

NaGISA data. For example, cover estimates can be done relatively quickly, and students, agencies, or local people can be trained to do so with high scientific accuracy. This opens opportunities to perform long-term monitoring at specific sites and/or the expansion of quantitative nearshore coverage with the inclusion of added manpower from local stakeholders. NaGISA's regionally organised network of nearshore researchers allows local scientists across the world to participate in this effort and thus make the final product larger than the sum of its individual parts.

NaGISA's specific scientific findings from its field surveys include inventories of marine flora and fauna, data on their abundance and biomass, new species records, species range extensions, habitat range extensions, biodiversity hot spots, and explanations of nearshore ecological processes and biodiversity drivers. All findings can now be analysed on regional as well as global scales.

Several new species were found and subsequently described during the NaGISA inventories. These new species discoveries included some small and inconspicuous species, like two cumaceans from the Gulf of Alaska (*Cumella oculatus* and *C. alaskensis*; Gerken 2009). These cumaceans are not only new species but their discovery was surprising as the genus Cumella is typically tropical rather than boreal–Arctic. Cumaceans, as filter feeders and surface deposit feeders, are ecologically important in energy transfer within the benthic food web and on the Alaskan shelf as they are important food for grey whales. Other, more conspicuous new species discovered were the golden V kelp in the Aleutian Islands, Alaska (*Aureophycus aleuticus*; Kawai et al. 2008). This kelp grows up to 3 m in length, and histological and genetic analyses show that it may not be closely related to other kelp species in the region. This opens interesting evolutionary and distributional questions about kelps in the North Pacific, where they form important habitats for associated biodiversity.

Another significant NaGISA accomplishment has been the discovery of the anomalodesmatan bivalve *Pholadomya candida* living in a *Thalassia testudinum* seagrass bed at Santa Marta, Colombia. This bivalve species belongs to the ancient family Pholadomyidae, a group of burrowing bivalves living on Earth since at least the Early Carboniferous (330 million years bp), which reached a high degree of diversification in Jurassic to Cretaceous times. *Pholadomya candida* had been collected alive only twice, with the last record in 1842, and, because living specimens had not been recorded for nearly 140 years, some authors considered the species extinct. The evolutionary implications of this re-discovery are remarkable. Comparative molecular sequencing of *P. candida* with other anomalodesmatan species and with representatives of other presumably related groups may provide clues of the evolution of the Anomalodesmata, as well as indications on the origin of the Myoida (Díaz et al. 2009).

Several new species distributional records and range extensions have been found during NaGISA sampling efforts. In the western Pacific, the solitary entoproct *Loxosomella sp.* was found in a seagrass sample from Akajima, Okinawa Prefecture, Japan. This is the first record of this animal group from sandy seagrass habitats in this region. Another interesting discovery was made in the Eastern Pacific with the coralline alga *Phymatolithon calcareum.* During NaGISA sampling, this species was found in its gametangial reproductive state (Konar et al. 2006). Although this species is relatively common and globally distributed, it was previously found only once in this reproductive state and that record was off the Atlantic coast of France (Mendoza & Cabioc'h 1998).

In the well-studied Cobscook Bay of the Gulf of Maine in the Atlantic Ocean region, NaGISA surveys found tens of benthic faunal taxa previously unreported from the area, from such diverse groups as hydrozoans (for example *Clytia gracilis*), mollusks (for example *Spisula solidissima, Astarte portlandica*), crustaceans (for example *Nebalia bipes, Metopella carinata*), polychaete worms (for example *Aricidea albatrossae, Euchone papillosa*), and bryozoans (for example *Haplota clavata, Cribrilina punctata*). Similarly, new species records for five macroalgal species were found at NaGISA sites in the Arctic Beaufort Sea, including the brown algae *Sphacelaria plumosa* and *S. arctica*, and the red algae *Rhodomela tenuissima* and *Scagelia cf americana.* Also at these sites, the common red alga *Phyllophora truncata* was often infested with what has been tentatively identified as an endophytic alga Chlorochytrium.

In addition to species-level discoveries, NaGISA also had some significant discoveries of habitat extensions. A major range extension was the discovery of a rhodolith habitat in the Eastern Pacific region (Konar et al. 2006). Rhodoliths are unattached calcareous red algae that form extensive beds, which provide habitat for many associated, sometimes commercially important species. Although rhodolith beds are widely distributed in temperate and tropical areas, the rhodolith bed discovery in Alaska's Prince William Sound in the North Pacific Ocean represents a significant northward extension of known rhodolith distribution. Also, in the Arctic Beaufort Sea, a new boulder field providing substrate for a diverse community of macroalgae and invertebrates was mapped in Camden Bay through NaGISA efforts (Iken & Konar 2007).

Regional comparisons have already yielded new insights into biodiversity patterns. Longitudinal comparisons in the Caribbean Seas region have shown that there is a decrease in species numbers from west to east. At the same time, this gradient in species numbers is not similarly reflected in the taxonomic structure of the communities (based on the index of taxonomic distinctiveness) as this is the same along that longitudinal gradient. The nested design of the NaGISA sampling protocol was used in the Gulf of Alaska in the EPAC region to analyse the contributions of local versus regional scales of variability in nearshore communities (Konar et al. 2009). Interestingly, most variability was associated with the local scale and very little with regional scales. On the local scale, the depth gradient was the most important factor contributing to variability, which was also found when only echinoderm distribution was analysed over the same spatial scales (Chenelot et al. 2007).

The number of species generally increases from the high intertidal to a depth of 1 m and then decreases with increasing subtidal depths. The large tidal range in the region effectively renders the 1 m depth stratum low intertidal and thus a suitable interface for a large variety of intertidal and subtidal organisms (Konar et al. 2009). Seasonal comparisons in the South American region (Puerto Madryn, Argentina) found that local biodiversity varies throughout an annual cycle in close relation to the presence of an invasive brown algal species (*Undaria pinnatifida*), which is sensitive to warm temperatures. During the austral winter, *U. pinnatifida* invades the rocky substrates replacing the natural community, but also attracts another community of gastropods, polychaetes, sea urchins, and other invertebrates that feed on the algae. As the water temperature increases in the austral summer, *U. pinnatifida* dies, and the natural community returns.

Along with reporting community patterns and biodiversity trends, it also is important to explain why and how these trends and patterns exist. Some research has already examined drivers of community patterns and biodiversity trends at various spatial scales (Coleman et al. 2006; Kuklinski et al. 2006; Scrosati & Heaven 2007; Wulff et al. 2009). NaGISA in the European Seas conducted an experimental study using a combination of long-term observations and field manipulations to show that rare species take advantage of environmental variability, becoming less rare in fluctuating environments (Benedetti-Cecchi et al. 2008). Hence, an increase in environmental variability, such as that expected under climate change models, may lead to major shifts in species composition within assemblages, with the prediction that currently rare species may become more dominant with increasing levels of environmental heterogeneity.

Remaining Questions

Although NaGISA efforts are greatly contributing to the field of nearshore biodiversity, many issues and questions remain. First and foremost, true estimates for global nearshore biodiversity do not exist. The million dollar question of how many organisms live in nearshore waters, still cannot be answered. It appears that the more regions that are sampled and the more taxonomists that are involved, the more new species and range extensions are found. We may never know exactly how many species live in the nearshore, but we can and should continue working towards increasingly accurate estimates.

Another question that still remains open is why certain areas have higher biodiversity (or abundances or biomass) than other areas. The more we learn about biodiversity trends and the physical and biological attributes that contribute to biodiversity hot spots, the easier it will be to answer the question of why these hot spots exist. From the data currently available, it appears that many of the hot spots occur because of various site-specific parameters (that is, hard substrate in an otherwise soft substrate environment, local oceanographic conditions). However, more information is needed to determine if and what biological and physical parameters will result in the existence of a hot spot and if large-scale generalisations of such relationships can be made.

Along with the questions, some problems also remain. One such problem that still exists in many regions is the surveying of remote and isolated areas. With the advances in remote sensing, these areas are becoming more accessible. The intertidal zone can be surveyed with remote sensing, using Ikonos satellite imaging, followed by hyperspectral imaging and ground-truthing (Larsen et al. 2009). In some areas, like the northern part of the Eastern Pacific region, programs exist that have already mapped nearshore coasts, such as the ShoreZone project. In many areas, the subtidal areas can be mapped and information such as bottom type and depth can be acquired from multibeam sonar acoustic mapping. This can then be ground-truthed with benthic sampling. This type of information will make discovering new biodiversity hot spots and describing patterns and processes in the nearshore much easier. Nevertheless, although such mapping efforts can supply guidance and large-scale coverage, the need for local ground-truthing and traditional establishment of biodiversity remains.

Although there has been much advancement in the knowledge of nearshore biodiversity, education in developing countries must continue. It has become evident that there is a need for expert services and facilities to process field samples efficiently and completely. Some regions have this service, such as the Atlantic Ocean region through the Atlantic Reference Centre (Huntsman Marine Science Centre, New Brunswick, Canada), which is in charge of processing, quality control/assurance, and archiving all Atlantic Ocean regional samples.

Current NaGISA data will culminate in 2010 with assessing spatial (for example latitudinal or longitudinal) trends of overall community patterns in rocky macroalgal systems and seagrass beds, as well as of selected taxonomic groups. Because of the relatively good taxonomic expertise available in most regions of the world, NaGISA is focusing in this first phase on patterns in macroalgae, polychaetes, gastropods, echinoderms, and decapods. However, there are many other taxonomic groups that are yet unexplored, but are no less ecologically important.

Not only do we know that biodiversity trends vary depending on the taxonomic group examined, but these patterns may be quite different for the rarer groups than the more common taxa. Similarly we have learned that biodiversity trends often depend on the depth strata examined, but without better knowledge of small-scale biodiversity patterns, overall trends will be difficult to determine. There are many questions that remain unanswered here, such as what is the global latitudinal trend for cnidarians, sponges, or bryozoans, and do these trends vary with depth.

Conclusions

NaGISA's major legacies thus far can be summarised as the following.

1. The creation of the first standardised global baseline of coastal biodiversity in rocky shores and seagrass beds from the intertidal zone to water depths of up to 20 m.
2. The establishment of a standardised sampling protocol that is suitable to analyse biodiversity trends on multiple spatial and temporal scales.

3. The improvement of benthic taxonomy.
4. The network of scientists and new scientific capacity-building around the world, a network that is now working together to address major questions in nearshore biodiversity.
5. The elucidation of the scales of temporal and spatial variability in nearshore habitats.
6. The addition of knowledge on the interactive effects of multiple drivers, including human activities, on spatial patterns of marine coastal biodiversity at the global scale.
7. The identification of hot spots of marine coastal biodiversity that can be suggested for new Marine Protected Areas.

The NaGISA project has sampled many sites throughout the world, but the efforts are still dwarfed compared with the vastness of the world's nearshore region. Some of the sampled sites have now been established for long-term monitoring. In all regions, there will and should be continued monitoring of selected NaGISA sites.

This monitoring will be done by a combination of researchers, elementary, high school, and university students, local communities, and other stakeholders.

NaGISA has particularly enhanced stakeholder "ownership" at many sites, similar to sponsorship of roadside clean-up programs. Although information for truly global comparisons is still lacking in many areas and for certain taxonomic groups, patterns in biodiversity are beginning to emerge.

More sites are continually being added and more taxonomists are being engaged. The momentum that NaGISA has started must continue if we are to get an increasingly accurate description of global diversity.

The NaGISA monitoring sites will assist with the identification of inter-annual variability. This is crucial to be able to distinguish short-term variability from longer-term changes that may be driven by climatic changes or anthropogenic pressures.

Such long-term changes will become measurable over time from NaGISA sites that are part of the long-term monitoring. In addition, the NaGISA–History of Marine Animal Populations collaboration, the History of the Nearshore (HNS) project, is identifying changes in nearshore communities that have occurred over decadal scales. By comparing historical baselines with present-day data, regional changes within the various HNS studies may be detected. Changes revealed by comparisons of several Atlantic HNS regions could, for example, produce a Pan-Atlantic pattern and identify driving factors. NaGISA has done much to not only advance the knowledge and appreciation of nearshore biodiversity, but it has started a momentum through its outreach, networking, and capacity building.

We may never be able to answer how many species live in the nearshore, but we will continue to produce a more accurate estimation and to explain why there are so many nearshore species and why they are distributed as they are.

4

Biological Research in Marine Life

The aim of this special issue of Biological Research is to present some of the latest advances in biological research of scientists under the Frontiers of Science Program of the Chilean Academy of Sciences of the Instituto de Chile. This program started 7 years ago, with the idea of incorporating into the Academy the very best young scientists working in Chile in order to discuss exciting advances and opportunities in their fields in a format that encourages informal collective, as well as one-on-one, discussions among participants. Several symposia have been organised over the years, and in each case speakers have been urged to focus their talks on describing current cutting-edge research in their disciplines to colleagues outside their field and to address questions such as: "What are the major research problems and distinctive tools of your field?" and "How might insights derived from other fields contribute to overcoming these limitations?" Participants remain active in the Frontiers of Science Program for an average of 3 years. Such groups have included leading researchers in disciplines such as astronomy, biology, chemistry, geology, mathematical sciences and physics. Here the Society of Biology has invited biologists that have been part of Frontiers of Science Groups to present their latest and most exciting work to the readers of Biological Research.

In the first study, Allende's team describes how a teleost, the zebrafish, can be useful to study the cellular and molecular mechanisms associated with copper toxicity in water, establishing a framework for further analyses of copper excess using an entire organism. The use of invertebrate models for research is also presented by Glavic et al. In this work, they describe how the stability of the Notch protein is dependent on the activity of GDP mannose dehydratase (GMD), and therefore on the levels of O-fucose moieties added to the Notch extracellular domain. These post-translational modifications will ultimately contribute to regulate the endocytosis and degradation of Notch.

The usefulness of vertebrate models for research is discussed by several papers describing immunology, cell biology, neurobiology and development. Thus, during the development of the nervous system morphogens are responsible for the establishment of neuronal territories. Palma, Larrain et al describe how the extracellular matrix may modulate the function of the morphogen Sonic Hedgehog (SHH). They specifically propose perlecan as the proteoglycan involved in both SHH localisation and activity, through a direct interaction among these molecules. After the establishment of neural territories, several cellular events lead to neuronal differentiation. Borquez and Gonzalez-Billault analyse the molecular mechanisms contributing to the establishing of cell polarity. This review presents evidence relating the ubiquitin proteosome and calpain systems, with cell migration and the development of an axon in neuronal cells, two processes that require the polarisation of cells. Another interesting aspect of neuronal function is described in the work of Couve's team. These authors review the state-of-the-art in studying the endoplasmic reticulum in neuronal cells, providing a detailed analysis of its structure, composition, dendritic distribution and dynamics. A functional link is provided with the analysis of the possible roles for the endoplasmic reticulum in synaptic transmission. In the immunological system, dendritic cells (DC) are responsible for maintaining adaptive immunity because they can modulate T-cell activities.

Kalergis' team present a review of the interactions between dendritic and T-cells. Consequences of altered DC-T cells are analysed in the context of autoimmune diseases. Pathological conditions are also presented in the paper by Marzolo and Farfan. This article deals with the role of megalin, a receptor for apolipoprotein E that is implicated in several diseases such as Lowe Syndrome, Dent Disease, Alzheimer's disease and gallstone disease. The authors review the current known functions and propose mechanisms that control megalin expression. Vegetal biology is also represented by the work of Moreno and Orellana, which describes the plant unfolded protein response (UPR). This is an up-to-date review showing the emergence of plant homologues for UPR proteins, and proposes some mechanistic linkage to physiological responses in plants.

Nature biologists are also represented in this special issue by three articles. Nespolo's team, analyse how morphological changes occurring during ontogeny may alter the metabolism of invertebrates. Metabolism in vertebrates is analysed by Ramirez-Otarola and Sabat. They present an article showing how dietary habits may influence the physiology of digestion in eight bird species. Changes in the enzymatic activities amongst species led the authors to propose a genetic base for such differences. As well, by analysing the distribution of breeding bird species, Abades and Marquet, characterise ecological phenomena. Based on statistical analyses of bird distribution, they propose that differences may be related to macroecological behaviour.

Finally, in a novel discipline, Gonzalez-Nilo et al. present a review of a new sub-discipline termed nanobioinformatics. The article shows some examples of nanobiotechnology in life sciences, and discusses new challenges for bioinformatics and computational chemistry.

We hope this number of the Biological Research *will be of interest for a broad number of biologists. We appreciate the energy and enthusiasm of the authors and other scientist that helped during the review process.*

Biological Research: Recent Advances in Angiotensin II

Angiotensin II (Ang II)* is a multifunctional hormone that influences the function of cardiovascular cells through a complex series of intracellular signalling events initiated by the interaction of Ang II with AT_1 and AT_2 receptors. AT_1 receptor activation leads to cell growth, vascular contraction, inflammatory responses and salt and water retention, whereas AT_2 receptors induce apoptosis, vasodilation and natriuresis. These effects are mediated via complex, interacting signalling pathways involving stimulation of PLC and Ca^{2+} mobilisation; activation of PLD, PLA_2, PKC, MAP kinases and NAD(P)H oxidase, and stimulation of gene transcription. In addition, Ang II activates many intracellular tyrosine kinases that play a role in growth signalling and inflammation, such as Src, Pyk2, p130Cas, FAK and JAK/STAT.

These events may be direct or indirect via transactivation of tyrosine kinase receptors, including PDGFR, EGFR and IGFR. Ang II induces a multitude of actions in various tissues, and the signalling events following occupancy and activation of Ang receptors are tightly controlled and extremely complex. Alterations of these highly regulated signalling pathways may be pivotal in structural and functional abnormalities that underlie pathological processes in cardiovascular diseases such as cardiac hypertrophy, hypertension and atherosclerosis.

Angiotensin II (Ang II) regulates blood pressure, plasma volume, sympathetic nervous activity and thirst responses. It also plays an important pathophysiological role in cardiovascular disease, including cardiac hypertrophy, myocardial infarction, hypertension and atherosclerosis. Ang II is produced systemically via the classical renin-angiotensin system and locally via the tissue renin-angiotensin system. Ang II was initially described as being primarily a vasoconstrictor peptide. However, recent studies demonstrate that Ang II has growth factor and cytokine-like properties as well.

At the cellular level Ang II modulates contraction, it regulates cell growth, apoptosis and differentiation, it influences cell migration and extracellular matrix deposition, it is pro-inflammatory, it stimulates production of other growth factors (e.g., platelet-derived growth factor, PDGF) and vasoconstrictors (e.g., ET-1), and it transactivates growth factor receptors (e.g., PDGFR, epidermal growth factor receptor (EGFR), and insulin-like growth factor receptor (IGFR)). The multiple actions of Ang II are mediated via specific, highly complex intracelular signalling pathways that are stimulated following initial binding of the peptide to its specific receptors. In mammalian cells Ang II binds to two distinct high-affinity plasma membrane receptors, AT_1 and AT_2.

The term "intracellular signalling pathway" includes the complex interrelated molecular cascades that transmit information from the membrane receptor to the intracellular proteins that regulate cell activities. The present review focuses on recent advances in Ang II signal transduction in cardiovascular cells. The growth factor and cytokine-like properties of this vasoconstrictor peptide will be highlighted and implications in cardiovascular disease will be discussed.

Angiotensin Receptors

AT_1 and AT_2 receptors, which are both seven transmembrane spanning G protein-coupled receptors, have been cloned and pharmacologically characterised. Pharmacologically the receptors can be distinguished according to inhibition by specific antagonists. AT_1 receptors are selectively antagonised by biphenylimidazoles such as losartan, whereas tetrahydroimidazopyridines such as PD123319 specifically inhibit AT_2 receptors (3,4). Two other Ang receptors have been described, AT_3 and AT_4 (5). However, the pharmacology of these receptors has not been fully characterised, and therefore these receptors are not yet included in a definitive classification of mammalian Ang receptors (6).

AT_1 Receptor

In humans there is only a single AT_1 receptor type, whereas in rodents, two subtypes of the AT_1 receptor have been identified (AT_{1a} and AT_{1b}). To date, AT_1 receptors have been shown to mediate most of the physiological actions of Ang II and this subtype is predominant in the control of Ang II-induced vascular functions (1). In the vasculature, AT_1 receptors are expressed mainly in smooth muscle cells (7). In the heart, AT_1 receptors are present in cardiomyocytes and fibroblasts (7). Ligand-receptor binding leads to activation of G proteins through exchange of GTP for GDP, resulting in the release of a and ßg complexes, which mediate downstream actions. AT_1 receptors interact with various heterotrimeric G proteins including Gq/11, Gi, Ga12 and Ga13. The different G protein isoforms couple to distinct signalling cascades. For example, Gq activation results in activation of phospholipase C (PLC), whereas GaI leads to cGMP formation. Although G protein-coupled receptors do not contain intrinsic kinase activity, they are phosphorylated on serine and threonine residues by members of the G protein receptor kinase family. AT_1 receptors are phosphorylated in the basal state and in response to Ang II stimulation. Various tyrosine kinases, including Janus kinases (JAK and TYK), Src family kinases, and focal adhesion kinase (FAK) can tyrosine phosphorylate AT_1 receptors (8).

AT_2 Receptor

The second major isoform of the Ang receptor, AT_2, is normally expressed at high levels in fetal tissues, and decreases rapidly after birth (9). In adults, AT_2 receptor expression is detectable in the pancreas, heart, kidney, adrenals, myometrium, ovary, brain and vasculature (9). In blood vessels, AT_2 receptors predominate in the adventitia and are detectable in the media. The AT_2 receptor is re-expressed in adults after vascular and cardiac injury and during wound healing and renal obstruction, suggesting a role for this receptor type in tissue remodelling, growth and/or development. The functional roles of AT_2 receptors are unclear, but these receptors may antagonize, under physiological conditions, AT_1-mediated effects by inhibiting cell growth, and by inducing apoptosis and vasodilation (10,11). Recently, in vivo studies of forearm blood flow in healthy human subjects found that the AT_2 receptor was not involved in modulating vascular tone in these subjects (12). Other studies suggest that AT_2 receptors also contribute to pathological processes associated with cardiac hypertrophy and inflammation (13). Signalling pathways through which AT_2 receptors mediate cardiovascular actions have recently been elucidated. Four major cascades are involved including 1) activation of protein phosphatases and protein dephosphorylation, 2) regulation of the nitric oxide-cGMP system, 3) stimulation of PLA_2 and release of arachidonic acid, and 4) sphingolipid-derived ceramide. These pathways have been reviewed elsewhere (14) and will not be discussed further here.

Intracellular Signals Induced by AT_1 Receptors

Ang II promotes its effects by acting directly through Ang II receptors, indirectly through the release of other factors, and via crosstalk with intracellular signalling pathways of other vasoactive agents, growth factors and cytokines. AT_1 receptors are coupled to multiple, specific signalling cascades, leading to diverse biological actions. The signalling processes are multiphasic with distinct temporal characteristics.

AT_1 Signalling Through Phospholipids

Phospholipase C. One of the earliest detectable events resulting from Ang II stimulation is a rapid, PLC-dependent hydrolysis of phosphatidylinositol-4, 5-bisphosphate (15). The PLC family includes three related enzymes: PLC-ß, PLC-g, and PLC-d which are regulated by either G proteins a and ßg (PLC-ß), by tyrosine phosphorylation (PLC-g) or by Ca^{2+} (PLC-d). Classically, AT_1 receptor activation results in a rapid production of 1,4,5-inositol triphosphate (IP_3), and a more sustained release of diacylglycerol, which are involved in Ca^{2+} mobilisation from the sarcoplasmic reticulum and stimulation of protein kinase C (PKC) (16), respectively. Ang II-stimulated IP_3 generation may also be mediated, in part, via tyrosine kinase-dependent pathways.

Increased intracellular Ca^{2+} results in vascular smooth muscle cell (VSMC) contraction, whereas PKC activation regulates intracellular pH through the Na^+/H^+ exchanger (15,16). PLC activation correlates temporally with initiation of contraction in isolated VSMC, as well as in intact small resistance arteries, and most likely constitutes the early signalling pathway for initiation of the Ca^{2+}-dependent, calmodulin activated phosphorylation of the myosin light chain which leads to cellular contraction. The cellular processes underlying Ang II-induced rise in $[Ca^{2+}]_i$ and pHi following AT_1 receptor stimulation have been extensively reviewed elsewhere (15,17).

Phospholipase D. Unlike PLC, which preferentially acts on phospholipids containing phosphoinositol, PLD hydrolyses phosphatidylcholine. The sustained activation of PLD is a major source of prolonged second messenger generation in VSMC and cardiomyocytes (18). Hydrolysis of phosphatidylcholine by PLD leads to the production of phosphatidic acid and subsequent generation of diacylglycerol by phosphatidic acid phosphohydrolase (18). Diacylglycerol is the physiological activator of PKC and is also a source of arachidonic acid. A number of PKC isoforms have been described including a, ß, d, e, μ, V, and l. Molecular mechanisms coupling AT_1 receptors to PLD involve Gßg and their associated Ga12 subunits, Src and RhoA (19). The downstream pathways associated with Ang II-induced activation of PLD in VSMC are PKC independent, but involve intracellular Ca^{2+} mobilisation and Ca^{2+} influx that is tyrosine kinase dependent. Ang II-induced PLD signalling has been implicated in cardiac hypertrophy, VSMC proliferation, and vascular contractility (20,21).

These actions are mediated via phosphatidic acid and other PLD metabolites, that influence vascular generation of superoxide anions by stimulating NAD(P)H oxidase, which activate tyrosine kinases and Raf and modulate intracellular Ca^{2+} signalling (18). The long-term signalling events associated with Ang II-stimulated growth and remodelling in the cardiovascular system are dependent, in large part, on PLD-mediated responses.

Phospholipase A_2. Ang II induces activation of PLA_2, which is responsible for release of arachidonic acid from cell membrane phospholipids (22). Released arachidonic acid is metabolised by cyclooxygenases, lipoxygenases or cytochrome P450 oxygenases to many different eicosanoids in vascular and renal tissues. Cyclooxygenases catalyse the formation of prostaglandin (PG) PGH_2, subsequently converted to thromboxane by thromboxane synthase, to PGI_2 (or prostacyclin) by prostacyclin synthase, or to PGE_2, PGD_2 or PGF_{2a}, by different enzymes (22). Lipoxygenases catalyse the formation of 5-, 12-, or 15-HPETEs, that then undergo spontaneous or peroxidase catalysed reduction to the corresponding HETEs and, in the case of 5-HPETE, to leukotrienes (22). Cytochrome P450 oxygenases catalyse arachidonic acid epoxidation to epoxyeicosatrieenoic acids, w and w-1 hydroxylation to 20- and 19-HETE, and allylic oxidation to other HETEs.

PLA_2-derived eicosanoids influence vascular and renal mechanisms important in blood pressure regulation. In VSMC and endothelial cells, these effects are mediated via AT_1 receptors, whereas in neonatal rat cardiac myocytes, neuronal cells and renal proximal tubule epithelial cells, Ang II-induced activation of PLA_2 occurs via AT_2 receptors (23). Ang II-elicited activation of vascular PLA_2 is dependent on $[Ca^{2+}]_i$, Ca^{2+} calmodulin dependent protein kinase II and mitogen-activated protein kinases (MAPK) (22,23). Activated PLA_2 and its metabolites in turn activate Ras/MAPK-dependent signalling pathways, amplifying PLA_2 activity and releasing additional arachidonic acid by a positive feedback mechanism. Ang II-generated eicosanoids regulate vascular contraction and growth, possibly by activating MAPK and redox-sensitive pathways. Thromboxanes are involved in Ang II-stimulated contraction, whereas vasorelaxant PGs such as PGE_2 and PGI_2 attenuate Ang II-mediated vasoconstriction in some vascular beds. Lipoxygenase derived eicosanoids also influence Ang II-elicited actions in VSMC. 12-HETE facilitates the stimulatory actions of Ang II on Ca^{2+} transients in cultured cells. Lipoxygenase inhibitors attenuate the vasoconstrictor action of Ang II and decrease blood pressure in SHR (24).

Phospholipid-derived second messengers resulting from Ang II-activated phospholipases. Shown are the major phospholipid-derived products from reactions catalysed by phospholipase A (PLA), phospholipase C (PLC) and phospholipase D (PLD). AT_1R = angiotensin 1 receptor, IP_3 = inositol triphosphate, DAG = diacylglycerol, PKC = protein kinase C.

AT_1-mediated Tyrosine Phosphorylation

A recent development in the field of Ang II signalling is the demonstration that AT_1 receptor activation is associated with increased protein tyrosine phosphorylation and activation of MAPK. These processes are characteristically associated with growth factors and cytokines. Accordingly, it is becoming increasingly evident that in addition to its potent vasoconstrictor properties, Ang II has mitogenic- and inflammatory-like characteristics. Ang II stimulates phosphorylation of many non-receptor tyrosine kinases including PLC-g, Src family kinases, JAK (JAK and TYK), FAK, Ca^{2+}-dependent tyrosine kinases (e.g., Pyk2), p130Cas and phosphatidylinositol 3-kinase (PI3K). In addition Ang II influences activity of receptor tyrosine kinases (RTK), such as EGFR, PDGFR and IGFR. The role of tyrosine kinases in Ang II-mediated signalling has been extensively reviewed (1,2,8). Only recent developments are discussed here.

Non-receptor Tyrosine Kinase Activation

Src family kinases. To date at least 14 Src-related kinases have been identified, of which the 60-kDa c-Src is the prototype. c-Src is abundantly expressed in vascular smooth muscle and is rapidly activated by Ang II in VSMC (25). Src plays an important role in Ang II-induced phosphorylation of PLC-g and IP_3 formation. Src, intracellular Ca^{2+} and PKC regulate Ang II-induced phosphorylation of p130Cas, a signalling molecule involved in integrin mediated cell adhesion.

Src has also been associated with Ang II-induced activation of Pyk2 and extracellular signal-regulated kinases (ERKs) as well as activation of other downstream proteins including pp120, p125Fak, paxillin, JAK2, signal transducers and activators of transcription 1 (STAT1), Ga, caveolin, and the adapter protein, Shc (26). Studies in VSMC isolated from human resistance arteries suggest that c-Src may also be important in the regulation of Ang II-stimulated Ca^{2+} mobilisation (27). Furthermore, c-Src mediates Ang II regulation of plasminogen activity in bovine aortic endothelial cells. Activation of c-Src is required for cytoskeletal reorganisation, focal adhesion formation, cell migration and cell growth (28). Increased activation of c-Src by Ang II may be an important mediator of altered VSMC function in hypertension.

Janus family kinases, tyrosine kinase and STAT activation. Similar to classical cytokine receptors, the AT_1 receptor stimulates JAK2 and TYK2, members of the JAK family . AT_1 receptor-induced activation of JAK leads to phosphorylation of the STAT proteins p91/84 (STAT1a/ ß), p113 (STAT2) and p92 (STAT3), which are transcription factors. Ang II-induced tyrosine phosphorylation and nuclear translocation of STAT1 require JAK2 and p59 Fyn kinase (a member of the Src family of kinases). p59 Fyn appears to act as a docking protein for both JAK2 and STAT1, which facilitates JAK2-mediated phosphorylation of STAT1.

JAK proteins are key mediators of mRNA expression and are characterised as "early growth response genes". JAK phosphorylates STAT proteins that are translocated to the nucleus, where they activate gene transcription. Electroporation of antibodies against STAT1 and STAT3 abolished VSMC proliferative responses to Ang II, but not to other growth factors, implicating an essential role of STAT proteins in Ang II-induced cell proliferation. The JAK-STAT signalling pathway activates early growth response genes, and may be a mechanism whereby Ang II influences vascular and cardiac growth, remodelling and repair.

Focal adhesion kinase and proline-rich tyrosine kinase 2. Ang II promotes cell migration and induces changes in cell shape and volume by activating FAK-dependent signalling pathways . Focal adhesion complexes, specialised sites of cell adhesion, act as supramolecular structures for the assembly of signal transduction mediators. The best-characterised tyrosine kinase localised to focal adhesion complexes is a 125-kDa protein, FAK. FAK exhibits extracellular matrix-dependent tyrosine autophosphorylation and physically associates with two non-RTK, c-Src and p59 Fyn (pp59), via their SH2 domains.

FAK autophosphorylation may also result in physical associations with PI3K, which is a 'downstream' tyrosine kinase involved in trophic cellular responses. As a consequence of its association with c-Src, FAK undergoes further tyrosine phosphorylation, which results in FAK binding to Grb2, an association with the GDP-GTP exchange protein, Sos, and Ras. This in turn leads to ERK1/2 activation. FAK is abundant in developing blood vessels, and elevation of its phosphotyrosine content in VSMC is a rapid response to Ang II.

Ang II-induced activation of FAK causes its translocation to sites of focal adhesion with the extracellular matrix and phosphorylation of paxillin and talin, which may be involved in the regulation of cell morphology and movement (32). AT_1-induced FAK activation also plays an important role in Ang II-mediated hypertrophic responses in VSMC. The link between the AT_1 receptor and FAK is unknown, but the Rho family of GTPases may be important.

Another FAK family member, Pyk2, also called cell adhesion kinase-ß, related adhesion focal tyrosine kinase and calcium-dependent tyrosine kinase (the rat homologue of Pyk2), is activated by AT_1 receptors and is dependent on increased intracellular Ca^{2+} (32).

Since Pyk2 is a candidate to regulate c-Src and to link G protein-coupled vasoconstrictor receptors with protein tyrosine kinase-mediated contractile, migratory and growth responses, it may be a potential point of convergence between Ca^{2+}-dependent signalling pathways and protein tyrosine kinase pathways in VSMC. In endothelial cells the balance of Pyk2 tyrosine phosphorylation in response to Ang II is controlled by Yes kinase (Src family kinase) and by a tyrosine phosphatase SHP-2 (33).

p130Cas. p130Cas is an Ang II-activated tyrosine kinase that plays a role in cytoskeletal rearrangement (34). This protein serves as an adapter molecule because it contains proline-rich domains, an SH3 domain, and binding motifs for the SH2 domains of Crk and Src.

p130Cas is important for integrin-mediated cell adhesion, by recruitment of cytoskeletal signalling molecules such as FAK, paxillin and tensin to the focal adhesions. In cultured VSMC, Ang II induces a transient increase in p130Cas tyrosine phosphorylation (28).

Some investigators have found this phosphorylation to be dependent on Ca^{2+}, c-Src and PKC, and to require an intact cytoskeletal network (28). Other studies reported that Ang II-induced activation of p130Cas is Ca^{2+} and PKC independent (35). Although the exact functional significance of Ang II-induced activation of p130Cas is unclear, it might regulate a-actin expression, cellular proliferation, migration and cell adhesion. p130Cas also plays a critical role in cardiovascular development and actin filament assembly.

Phosphatidylinositol 3-kinase. PI3K is a heterodimeric enzyme composed of a p85 adapter and a p110 catalytic subunit. PI3K catalyses the synthesis of 3-phosphorylated phosphoinositides. The major products of PI3K influence cell survival, metabolism, cytoskeletal reorganisation and membrane trafficking and have recently been identified to play an important role in the regulation of VSMC growth (36). PI3K, characteristically associated with tyrosine kinase receptors, is also activated by AT_1 receptors (37,38). In VSMC Ang II stimulates activity, phosphorylation and migration of PI3K, and induces translocation of the p85 subunit from the perinuclear area to foci throughout the cytoplasm and the cytoskeletal apparatus (37).

PI3K inhibition by wortmannin and LY294002 blocks Ang II-stimulated hyperplasia in cultured rat cells, suggesting the important regulatory role of this non-RTK in VSMC growth (37). Several molecular targets for PI3K have been identified, including centaurin, the actin-binding-protein profilin, phosphoinositide dependent kinases, the atypical PKCs, PLC-g, Rac1, c-Jun N-terminal protein kinase (JNK) and the protein Ser/Thr kinase (Akt)/protein kinase B (PKB) (38). Akt/PKB has recently been identified as an important PI3K downstream target in Ang II-activated VSMC. It regulates protein synthesis by activating p70 S6-kinase and it modulates Ang II-mediated Ca^{2+} responses in aortic cells by stimulating Ca^{2+} channel currents. Akt/PKB has also been implicated to protect VSMC from apoptosis and to promote cell survival by influencing Bcl-2 and c-Myc expression and by inhibiting caspases (38). Mechanisms whereby the AT_1 receptor mediates activation of PI3K-dependent Akt/PKB are unclear, but redox-sensitive pathways and c-Src may be important. Although the exact role of PI3K in Ang II signalling in VSMC has not yet been established, it is possible that this complex pathway may control the balance between mitogenesis and apoptosis.

Tyrosine kinase pathways stimulated by angiotensin II (Ang II) in vascular smooth muscle cells. Ang II activates Src, which regulates phospholipase C-g (PLC-g)- and extracellular signal-regulated kinase (ERK)-dependent signalling pathways. Ang II binding to the angiotensin 1 receptor (AT_1R) induces the physical association and activation of JAK2/TYK2 (Janus kinase/s) as indicated by the dashed line. JAK2/TYK2 phosphorylates STAT proteins (signal transducers and activators of transcription) that are translocated to the nucleus where they activate gene transcription. Ang II also

Receptor Tyrosine Kinases

Increasing evidence suggests that mitogenic responses to AT_1 receptor activation may be mediated by activation of RTK. Ang II can activate RTK, even though it does not directly bind to RTK. This process of transactivation has been demonstrated for EGFR, PDGFR and IGFR and has recently been reviewed . Mechanisms underlying Ang II-induced transactivation of RTKs include activation of tyrosine kinases (Pyk2 and Src) and redox-sensitive processes . EGFR transactivation seems to be a Ca^{2+}-dependent process, whereas PDGFR transactivation is Ca^{2+} independent. Recently, a novel concept for EGFR transactivation by G protein-coupled receptors was suggested.

In response to G protein-coupled receptor agonists (endothelin-1, thrombin, carbachol, lysophosphatidic acid and tetradecanoyl phorbol-13, acetate), heparin binding-EGF is generated by cleavage of pro-heparin-binding-EGF by metalloproteinase (40). Free heparin-binding-EGF then binds to EGFR resulting in EGFR homodimerisation and autophosphorylation. Similar processes have been demonstrated for IGFR transactivation. The role of these mechanisms in RTK transactivation by Ang II is unclear as some studies failed to demonstrate a role for metalloproteinase in AT_1-mediated EGFR transactivation in VSMC and cardiac fibroblasts .

In vitro evidence suggests that AT_1 receptor-induced EGFR transactivation is important for some of the trophic effects of Ang II. For example, AT_1 receptor-elicited tyrosine phosphorylation and activation of EGFR stimulated downstream activation of ERK1/2 and VSMC hyperplasia . In rat VSMC, both Ang II-induced nuclear proto-oncogene expression and increase in c-Fos protein were prevented by treatment with EGFR kinase inhibitor. Ang II-mediated EGFR transactivation also plays a role in p70 ribosomal protein S6-kinase induced protein synthesis. Furthermore, recent studies have demonstrated that EGFR activation is involved in Ang II-induced vascular contraction.

Mitogen-activated Protein Kinases

Mitogen-activated protein kinases are a family of serine/threonine protein kinases that mediate nuclear transduction of extracellular signals by intracellular protein phosphorylation, leading to a cascade of transcription factor activation, enhanced gene expression and trophic cellular responses. Mammalian MAPKs are grouped into six major subfamilies: a) ERK1/2 (also known as p42-kDa MAPK and p44-kDa MAPK, respectively), b) JNK/stress-activated protein kinases (JNK/SAPK), c) p38 MAPK, d) ERK6, p38-like MAPK, e) ERK3, and f) ERK5 (also called Big MAPK 1). MAPK-dependent signalling pathways have been associated with cellular growth and apoptosis, cellular differentiation and transformation and vascular contraction. ERK1/2 is activated in response to growth and differentiation factors, whereas JNKs and p38 MAPK are usually activated in response to inflammatory cytokines and cellular stress.

Ang II differentially activates the three major members of the MAPK family, ERK1/2, JNKs and p38 MAPK (43,44). Induction of MAPK activation typically involves phosphorylation by a MAPK kinase, also known as MEK. MEK is, in turn, regulated by other MEK kinases, including Raf-1. Although activated by similar stimuli, the signalling processes leading to JNK and p38 MAPK activation are quite different. The best characterised MAPK cascade is the Raf-Ras-MEK-ERK1/2 pathway.

Events downstream to MAPK activation are numerous and heterogeneous and include PLA_2, cytoskeletal proteins, the MAPK-activated protein kinase 2, and the $pp90^{rsk}$ protein kinase, which can translocate to the nucleus and activate transcription factors . Once phosphorylated, ERKs translocate to the nucleus to phosphorylate transcription factors and thereby regulate gene expression of cell cycle-related proteins. In VSMC, another downstream target of ERK is the serine/threonine protein kinase $pp90^{rsk}$, which phosphorylates the S6 ribosomal protein and stimulates protein synthesis. ERK1/2 activation ultimately results in enhanced proto-oncogene expression, and activation of the activator protein 1 complex transcription factor and probably regulates cell cycle progression as well as protein synthesis in VSMC. Ang II may also induce protein synthesis by an ERK-independent pathway in part via activation of the 70-kDa S6-kinase.

Other downstream targets of MAPK include cyclooxygenase-2, the contractile regulatory protein h-caldesmon, the high-molecular weight form of caldesmon, myelin basic protein, microtubule-associated protein, Ca^{2+} channels and the Na^{+}/H^{+} exchanger . The functional outcome of MAPK activation probably depends in part on the availability of downstream substrates.

Ang II activates the MAPK signalling cascade at various intracellular levels. It induces tyrosine and threonine phosphorylation of ERK1/2, JNK/SAPK and p38 MAPK in cultured VSMC, as well as in intact arteries. It stimulates phosphorylation of Ras, Raf and Shc and it increases activity of MEK kinase and MEK. In addition, Ang II increases activation of vascular Src and Pyk2, potential links between the Ang receptor and ERK. Ang II-stimulated ERK1/2 is associated with increased expression of the early response genes c-*fos,* c-*myc* and c-*jun,* DNA synthesis, cell growth and differentiation and cytoskeletal organisation. In addition to ERKs, Ang II activates JNK/SAPKs, which regulate VSMC growth by promoting apoptosis or by inhibiting growth . Ang II induces phosphorylation of JNK/SAPK via p21 activated kinase (aPAK) which is dependent on intracellular Ca^{2+} mobilisation and on PKC activation. Following phosphorylation, the isoforms JNK-1 and JNK-2 translocate to the nucleus to activate transcription factors, such as c-Jun, ATF-2 and Elk-1 (46). Ang II appears to activate VSMC ERK1/2 and JNK/SAPK via different signalling pathways. ERK phosphorylation occurs via a Ca^{2+}-dependent or -independent pathway that involves c-Src and the atypical PKC isoform PKC-x, whereas JNK/SAPK activation occurs via a Ca^{2+}-dependent pathway that involves a tyrosine kinase other than Src and a novel PKC isoform . Furthermore, whereas Ang II-induced phosphorylation peaks within 5 min, kinase activation is maximal at about 30 min. The functional effects of Ang II-induced signalling of ERK1/2 and JNK/SAPK in VSMC probably relate to regulation of cell growth. Ang II-activated ERK1/2 and JNK/SAPKs have opposite growth effects, with ERK1/2 being facilitative and JNK/SAPK inhibitory. These signalling processes and associated cellular functions are important in vascular damage associated with cardiovascular disease.

Ang II also phosphorylates vascular p38 MAPK, which plays an important role in inflammatory responses, apoptosis and inhibition of cell growth (48). The p38 MAPK pathway has been implicated in various pathological conditions including cardiac ischemia, ischemia/ reperfusion injury, cardiac hypertrophy, progression of atherosclerosis and arterial remodelling in hypertension.

The specific upstream and downstream regulators of Ang II-activated p38 in VSMC are unclear, but p38 MAPK could be a negative regulator of ERK1/2 (48). p38 MAPK has been implicated to be an essential component of the redox-sensitive signalling pathways in Ang II-activated VSMC (48).

Inactivation of Ang II-stimulated MAPK occurs via MAPK phosphatase 1 (MKP-1) induced dephosphorylation of both tyrosine and threonine on MAPK. Inhibition of MKP-1 results in sustained activation of MAPK in response to Ang II, suggesting that this enzyme is primarily responsible for the termination of the MAPK signal (49).

Ang II induces activation of MKP-1, as well as tyrosine phosphatase (PTP-1C), and Ser/Thr phosphatase PP2A (49). These effects appear to be mediated via the AT_2 receptor subtype, which has been associated with inhibition of cell growth and apoptosis. Accordingly, AT_1 receptors induce growth via stimulation of ERK-dependent signalling pathways, whereas AT_2 receptors oppose these effects by stimulating MKP-1 activity to inhibit ERK activity, and to arrest the cell growth signal. Termination of Ang II-stimulated MAPK activity may also involve activation of PKA, which inhibits the phosphorylation of Raf-1.

Ang II-stimulated extracellular signal-regulated kinase (ERK)-dependent signalling pathways in vascular smooth muscle cells. The ERK phosphorylation cascade is initiated by Ang II binding to angiotensin 1 receptors (AT_1R) that induce Shc-Grb2-Sos formation (tyrosine phosphorylation of Shc) which in turn facilitates guanine nucleotide exchange on the small G protein Ras-GDP/GTP.

Activated Ras-GTP interacts with the Ser/Thr kinase Raf (mitogen-activated protein kinase (MAPK) kinase kinase (MAPKKK)) which translocates to the cell membrane. Activation of Raf leads to phosphorylation of two serine residues present in MEK (MAPK/ERK kinase) which in turn phosphorylates Thr/Tyr and activates MAPK, present as a 44- (ERK-1) and a 42-kDa isoform (ERK-2). Phosphorylated ERK has diverse intracellular protein targets which it phosphorylates and activates. Dephosphorylation of ERK is accomplished by activation of MAPK phosphatase-1 (MKP-1). PLA_2 - phospholipase A_2.

Small G Proteins (Rho family)

In addition to signalling through heterotrimeric G proteins, recent evidence suggests that AT_1 receptors activate monomeric small (21 kDa) guanine nucleotide-binding proteins (small G proteins) in VSMC. The small G protein superfamily comprises five subfamilies (Ras, Rho, ADP ribosylation factors, Rab and Ran) that act as molecular switches to regulate cellular responses (50). Of these, the Rho subfamily (RhoA, Racl and Cdc42) has been associated with Ang II signalling. In its GDP-bound form, RhoA is inactive and is mainly cytoplasmic. RhoA is activated upon the exchange of bound GDP for GTP and for its activation, requires post-translational modification (geranyl-geranylation) (50). One of the major downstream targets of RhoA is Rho kinase, which promotes VSMC contraction via the phosphorylation of the myosin-binding subunit of myosin light chain phosphatase, thereby inhibiting phosphatase activity (51). This contributes to increased Ca^{2+} sensitisation. Various vasoactive agonists signalling through G protein-coupled receptors enhance the sensitivity of the contractile machinery to changes in $[Ca^{2+}]_i$ via RhoA/Rho kinase. Whether this system contributes to contraction following AT_1 receptor activation awaits clarification. However, RhoA seems to play a role in AT_1-stimulated growth signalling pathways and in processes associated with atherosclerosis. Yamakawa et al. (52) demonstrated in rat VSMC that Ang II increases c-fos expression and protein synthesis via RhoA/Rho kinase-dependent mechanisms that do not involve ERK1/2 or p70 S6-kinase activation. Funakoshi et al. (53) reported that Rho kinase mediates Ang II-induced monocyte chemoattractant protein-1 expression in rat VSMC. RhoA has also been shown to be important in AT_1-mediated PLD activation.

Ang II also activates Racl, another small G protein. Racl participates in cytoskeletal organisation, cell growth, inflammation and regulation of NAD(P)H oxidase (50). In VSMC, Ang II activates Racl, which is an upstream regulator of p21-activated kinase and JNK. Racl also plays a role in Ang II-induction of gene transcription and in the regulation of NAD(P)H oxidase, which mediates generation of superoxide anions ($.O_2^-$) in vascular cells (50).

Rho-regulated signalling pathways by angiotensin II (Ang II). Rho is activated through AT_1 receptors. Rho regulates the activity of many signalling pathways that influence contraction, growth and cytoskeletal organisation. MBS, myosin binding subunit of the myosin light-chain phosphatase; PRK2, protein kinase C-related kinase; SRF, serum response factor

Generation of Reactive Oxygen Species

Reactive oxygen species such as O_2^- and hydrogen peroxide (H_2O_2) act as intercellular and intracellular second messengers that may play a physiological role in vascular tone and cell growth, and a pathophysiological role in inflammation, ischemia-reperfusion, hypertension and atherosclerosis. Xanthine oxidase, mitochondrial oxidases and arachidonic acid are the major sources of oxidative molecules in non-vascular tissue, whereas a nonmitochondrial, membrane-associated NAD(P)H oxidase appears to be the most important source of $.O_2^-$ in vascular cells (54). This enzyme transfers electrons from NADH or NADPH to molecular oxygen, producing $.O_2^-$. The complete molecular structure of the vascular oxidase is unknown, but it shares some features with the neutrophil oxidase, which comprises five subunits: a 22-kDa a-subunit ($p22^{phox}$), a glycosylated 91-kDa ß-subunit ($gp91^{phox}$), which together make up cytochrome b_{558}, the electron transfer element; cytosolic components $p47^{phox}$ and $p67^{phox}$, and a low-molecular weight G protein, Rac1 or Rac2. Recent studies by Rueckschloss et al. (55) demonstrated complete homology of the cDNA sequence of $gp91^{phox}$ in human umbilical vein endothelial cells and neutrophils.

Alternatively, differences in the levels of gene expression of $gp91^{phox}$ and $p67^{phox}$ were observed, with human umbilical vein endothelial cells exhibiting 1.1 and 2.5%, respectively, of the levels of expression of these genes, compared with those of neutrophils. By contrast, gene expression for $p22^{phox}$ and $p67^{phox}$ was comparable between these two cell types. These observations suggest that the levels of expression of $gp91^{phox}$ and $p67^{phox}$ genes may be rate-limiting factors for human vascular cell $.O_2^-$ production. Upon activation, the $p47^{phox}$ and $p67^{phox}$ proteins are translocated to the membrane and associate with the cytochrome b_{558}, creating the active oxidase. In VSMC, Ang II increases $.O_2^-$ production by activating NAD(P)H oxidase. This effect is sustained and probably contributes to long-term signalling events such as cell growth.

Generation of reactive oxygen species is regulated by various cytokines and growth factors, including Ang II, which increases $.O_2^-$ and H_2O_2 production in cardiac, vascular smooth muscle, endothelial, adventitial and mesangial cells (21,54,56). Increased production of reactive oxygen species has been implicated in the pathogenesis of Ang II-induced but not catecholamine-induced hypertension (57). Mechanisms underlying oxidative stress-induced hypertension may be associated with the vascular mitogenic effects of $.O_2^-$ and H_2O_2, decreased bioavailability of endothelium-derived nitric oxide and contractile actions of $.O_2^-$ and H_2O_2. Growth of VSMC by Ang II has an essential redox-sensitive component, which appears to be mediated in part via activation of p38 MAPK and JNK (48). Another redox-sensitive process cascade whereby Ang II influences cell function is through phosphorylation of the cell survival protein kinase Akt/PKB and activation of NF-kB, important in inflammatory responses (54). Generation of reactive oxygen species in the vasculature. Many enzyme systems stimulate production of superoxide anion (·O_2^-) from O_2. These include NADH/NADPH oxidase, xanthine oxidase, lipoxygenase, cyclooxygenase, P450 mono-oxygenase and mitochondrial oxidative phosphorylation. NADH/NADPH oxidase is a multisubunit enzyme that is the major regulated source of reactive oxygen species in endothelial and vascular smooth muscle cells. Dismutation of ·O_2^- spontaneously or enzymatically by superoxide dismutase (SOD) produces hydrogen peroxide (H_2O_2) that can undergo further reactions to generate the highly reactive hydroxyl radical (·OH). H_2O_2 may be metabolised by catalase or peroxidases to H_2O and O_2. Downstream targets of ·O_2^- and H_2O_2 include ERK5, p38, tyrosine kinases and NF-kB. MAPK = mitogen-activated protein kinase; ERK = extracellular signal-regulated kinase; Akt/PKB = protein Ser/Thr kinase

Ang II-induced Expression of Proto-oncogenes and Growth Factors

Long-term control of Ang II-regulated cellular growth, adhesion, migration, fibrosis and collagen deposition within the vasculature involves protein synthesis. Ang II induces the expression of several proto-oncogenes in human and rat VSMC, including c-*fos*, c-*jun*, c-*myc*, *erg-1*, VL-30, and proto-oncogene/activator protein 1 complex (58). Stimulation of early response genes by Ang II is associated with increased gene expression and production of growth factors, such as PDGF, EGF, transforming growth factor-ß, IGF-1, basic fibroblast growth factor and platelet activating factor, vasoconstrictor agents, such as ET-1, adhesion molecules such as ICAM-1, VCAM-1 and E-selectin, and integrins anß3 and ß5, and chemotactic factors such as tumour necrosis factor-a and monocyte chemoattractant protein-1 (58). These agents may contribute indirectly to the trophic and inflammatory actions of Ang II in cardiovascular tissue.

Ang II influences the architecture and integrity of the vascular wall by modulating cell growth as well as regulating extracellular matrix composition. Ang II increases expression and production of fibronectin, collagen type 1, tenascin, glycosaminoglycans, chondroitin/dermatan sulfates and proteoglycans, major constituents of the extracellular matrix in the vessel wall. In VSMC, mesangial cells and endothelial cells, Ang II increases levels and activity of plasminogen activator inhibitor-1, influencing fibrinolysis, extracellular matrix turnover and degradation and regulation of cell migration. Some of these actions have been linked to the AT_4 receptor subtype. However, this remains to be clarified. Ang II also stimulates the activity of matrix metalloproteinases (58) responsible for extracellular matrix degradation. Accordingly, Ang II influences vascular structure by stimulating synthesis of structural components of the extracellular matrix and by increasing production of factors that degrade the extracellular matrix proteins.

Conclusions

Ang II activates multiple signalling pathways. Until recently, the signalling events elicited by Ang II were considered to be rapid, short-lived and divided into separate linear pathways. However, these major intracellular signalling cascades do not function independently and are actively engaged in crosstalk. Downstream signals from the Ang II-bound receptors converge to elicit complex and multiple responses (reviewed in 59,60). Ang II induces a multitude of actions in various tissues, and the signalling events following occupancy and activation of angiotensin receptors are tightly controlled and extremely complex. Furthermore, the fact that Ang II transactivates EGFR, PDGF and IGFR suggests that RTKs are downstream targets of AT_1 receptors. Alterations of these highly regulated signalling pathways in cardiovascular cells may be pivotal in structural and functional abnormalities that underlie vascular pathological processes such as hypertension and atherosclerosis. Although there has been significant progress in the last few years in the elucidation of aberrations in Ang II-induced signal transduction in hypertension, we still know very little about the processes that underlie these phenomena, and at what point some pathways become more important than others. With the availability today of molecular and pharmacological tools that allow manipulation of specific signalling molecules, identification of distinct abnormalities in intracellular signalling should be possible. This will further our understanding of the role of Ang II in physiological and pathophysiological processes and will help define novel therapeutic targets in cardiovascular disease.

5

New Discoveries in Sea

Hydrogen-Powered Bacteria Discovered In Mussels

Researchers searching for new ways to make hydrogen a viable part of our energy future have a new place to look; a natural example of a living hydrogen-powered 'fuel cell'. During a recent expedition to hydrothermal vents in the deep sea, researchers from the Max Planck Institute of Marne Microbiology and the Cluster of Excellence MARUM discovered mussels that have their own on-board fuel cells in the form of symbiotic bacteria that use hydrogen as an energy source. Their results suggest that the ability to use hydrogen as a source of energy is widespread in hydrothermal vent symbioses.

Deep-sea hydrothermal vents are formed at mid-ocean spreading centres where tectonic plates drift apart and new oceanic crust is created by magma rising from deep within the Earth. When seawater interacts with hot rock and rising magma, it becomes superheated, dissolving minerals out of the Earth's crust. At hydrotermal vents, this superheated energy-laden seawater gushes back out into the ocean at temperatures of up to 400 degrees Celsius, forming black smoker chimneys where it comes into contact with cold deep-sea water. These hot fluids deliver inorganic compounds such as hydrogen sulfide, ammonium, methane, iron and hydrogen to the oceans. The organisms living at hydrothermal vents oxidise these inorganic compounds to gain the energy needed to create organic matter from carbon dioxide. Unlike on land, where sunlight provides the energy for photosynthesis, in the dark depths of the sea, inorganic chemicals provide energy for life in a process called chemosynthesis. When hydrotheral vents were first discovered more than 30 years ago, researchers were astounded to find that they were inhabited by lush communities of animals such as worms, mollusks and crustaceans, most of which were completely unknown to science. The first to investigate these animals quickly realised that the key to their survival was their symbiotic association with chemosynthetic microbes, which are the on-board power plants for hydrothermal vent animals. Until now, only two sources of energy were known to power chemosynthesis by symbiotic bacteria at hydrothermal vents: Hydrogen sulfide, used by sulfur-oxidising symbionts, and methane, used by methane-oxidising symbionts. “We have now discovered a third energy source” says Nicole Dubilier from the Max Planck Institute of Marin Microbiology in Bremen, who led the team responsible for this discovery.

The discovery began at the Logatchev hydrothermal vent field, at 3000 m depth on the Mid-Atlantic Ridge, an undersea mountain range halfway between the Caribbean and the Cape Verde Islands. The highest hydrogen concentrations ever measured at hydrothermal vents were recorded during a series of research expeditions to Logatchev. According to Jillian Petersen, a researcher with Nicole Dubilier, “our calculations show that at this hydrothermal vent, hydrogen oxidation could deliver seven times more energy than methane oxidation, and up to 18 times more energy than sulfide oxidation”.

In the gills of the deep-sea mussel Bathymodiolus puteoserpentis, one of the most abundant animals at Logatchev, the researchers discovered a sulfur-oxidising symbiont that can also use hydrogen as an energy source. To track down these hydrogen-powered on-board ‘fuel cells’ in the deep-sea mussels, the researchers deployed two deep-sea submersibles, MARUM-QUEST from MARUM at the University of Bremen, and KIEL 6000 from IFM-GEOMAR in Kiel. With the help of these remotely-driven submersibles, they sampled mussels from sites kilometres below the sea surface. Their shipboard experiments with live samples showed that the mussels consumed hydrogen. Once the samples were back in the laboratory on land, they were able to identify the mussel symbiont hydrogenase, the key enzyme for hydrogen oxidation, using molecular techniques.

The mussel beds at Logatchev form a teeming expanse that covers hundreds of square metrees and contains an estimated half a million mussels. “Our experiments show that this mussel population could consume up to 5000 litres of hydrogen per hour” according to Frank Zielinski, a former doctoral student in Nicole Dubilier’s Group in

Bremen, who now works as a post-doctoral researcher at the Helmholtz Centre for Environmental Research in Leipzig. The deep-sea mussel symbionts therefore play a substantial role as the primary producers responsible for transforming geofuels to biomass in these habitats. "The hydrothermal vents along the mid-ocean ridges that emit large amounts of hydrogen can therefore be likened to a hydrogen highway with fuelling stations for symbiotic primary production" says Jillian Petersen.

Even the symbionts of other hydrothermal vent animals such as the giant tubeworm Riftia pachyptila and the shrimp Rimicaris exoculata have the key gene for hydrogen oxidation, but remarkably, this had not been previously recognised. "The ability to use hydrogen as an energy source seems to be widespread in these symbioses, even at hydrothermal vent sites with low amounts of hydrogen" says Nicole Dubilier.

Deep in The Ocean, A Clam that Acts Like a Plant

How does life survive in the black depths of the ocean? At the surface, sunlight allows green plants to "fix" carbon from the air to build their bodies. Around hydrothermal vents deep in the ocean live communities of giant clams with no gut and no functional digestive system, depending on symbiotic bacteria to use energy locked up in hydrogen sulfide to replace sunlight. Now, the genome of this symbiont has been completely sequenced and published in Science.

"The difference here is that while plants get their energy and carbon via photosynthesis by chloroplast symbionts, this clam gets its energy via chemosynthesis," said Jonathan Eisen, a professor at the UC Davis Genome Centre and an author on the paper.

The actual work of photosynthesis in green plants is done by chloroplasts, descended from primitive single-celled organisms that were incorporated into other cells billions of years ago.

"The energy from hydrogen sulfide is used to drive carbon fixation in much the same way that chloroplasts carry out carbon fixation," Eisen said. The symbiotic bacteria also fix nitrogen and produce amino acids, vitamins and other nutrients required by the clam.

Studies of the deep sea have implications for studying the origins of life, Eisen said. Life on Earth may have got its start with microbes living on such chemical reactions, before the evolution of photosynthesis.

"And they're just plain interesting," Eisen added.

If you were thinking that giant clams sound tasty, think again. The hydrogen sulfide gives them a strong smell of rotting eggs.

Mysterious New Algae Forces Rethink In Ocean's Carbon And Nitrogen Cycles

An unusual microorganism discovered in the open ocean may force scientists to rethink their understanding of how carbon and nitrogen cycle through ocean ecosystems.

A research team led by Jonathan Zehr, a marine scientist at the University of California, Santa Cruz, characterised the new microbe by analysing its genetic material, even though researchers have not been able to grow it in the laboratory. Zehr said the newly described organism seems to be an atypical member of the cyanobacteria, a group of photosynthetic bacteria formerly known as blue-green algae.

"This research has revealed a big surprise about th microbiology of the oceans, and the complex integration of the ocean's nitrogen and carbon cycles," said Phillip Taylor, section head in the National Science Foundation (NSF)'s Division of Ocean Sciences, which funded the work. "The fact that nitrogen fixation in these abundant unicells is decoupled from photosynthesis is intriguing," said Taylor. "This unique adaptation brings up questions about the role of these abundant microbes in the ocean."

Unlike all other known free-living cyanobacteria, this one lacks some of the genes needed to carry out photosynthesis, the process by which plants use light energy to make sugars out of carbon dioxide and water.

The mysterious microbe can do something very important, though: It provides natural fertilizer to the oceans by "fixing" nitrogen from the atmosphere into a form useable by other organisms.

"For it to have such an unusual metabolism is very exciting," Zehr said. "We're trying to understand how something like this can live and grow with so many missing parts."

Earlier research by Zehr's group had revealed surprisingly large numbers of novel nitrogen-fixing cyanobacteria, including the one that is the focus of this study, in the open ocean.

Although 80 percent of Earth's atmosphere is nitrogen, most organisms cannot use it unless it is "fixed" to other elements to make molecules like ammonia and nitrate. Because nitrogen is essential for all forms of life, nitrogen fixation is a major factor controlling overall biological productivity in the oceans.

The new microbe is one of the most abundant nitrogen fixers in many parts of the ocean, Zehr said.

"I had begun to suspect that there was something missing in this organism's genome, and the genome sequencing confirmed that," said Zehr.

The results showed that it is missing the entire set of genes needed for photosystem II and carbon fixation, essential parts of the molecular machinery that carries out photosynthesis in plants and cyanobacteria.

"That has multiple implications," Zehr said. "It must have a 'lifestyle' that's very different from other cyanobacteria. Ecologically, it's important to understand its role in the ecosystem and how it affects the balance of carbon and nitrogen in the ocean."

During photosynthesis, photosystem II generates oxygen by splitting water molecules. Because oxygen inhibits nitrogen fixation, most nitrogen-fixing cyanobacteria only fix nitrogen at night, or do it in specialised cells. The lack of photosystem II enables the new microbe to fix nitrogen during the day, Zehr said.

But without photosynthesis, it can't take carbon dioxide from the atmosphere and convert it into sugars. So it's not clear how the new microbe feeds itself. Either it has some way of feeding on organic matter in its environment, or it lives in close association with other organisms that provide it with food, Zehr said.

"It would make a perfect symbiont because it could feed nitrogen to its host and live on the carbon provided by the host," he said.

Photosystem II is large complex of multiple proteins and chlorophyll molecules, but the team was unable to find any of the genes for the photosystem II core proteins. The genes for photosystem I appeared in the sequencing data, as did genes for both photosystems from the small numbers of contaminating cyanobacteria in the sample.

Zehr said he plans to continue research on the new microbe and fill some gaps in the present knowledge and efforts are currently underway to map the microbe's presence in the oceans and determine its global abundance.

Zehr is also interested in how its metabolism differs from other known cyanobacteria. If it can be cultured, there may be ways to exploit this organism's unusual metabolism i biotechnology applications, he said.

New DNA sequencing technology provided by 454 Life Sciences enabled rapid sequencing of the organism's genome. Zehr's coauthors on the paper include graduate student Shelley Bench, researcher Brandon Carter, and postdoctoral scholars Ian Hewson, Tuo Shi, and James Tripp of UCSC, as well as Faheem Niazi and Jason Affourtit of 454 Life Sciences.

The work was also supported by the Gordon and Betty Moore Foundation; sequencing was provided by 454 Life Sciences. A paper describing the new findings appears in the November 14 issue of the journal *Science*.

Sea Slugs Generating Green Energy

Photosynthesizing sea slugs take 'you are what you eat' to an extreme: by eating photosynthesizing algae, these "solar-powered" sea slugs are able to live off photosynthesis for months. How does this work? Is this just a straightforward case of symbiosis between algae and sea slugs?

It turns out that this is not a case of symbiosis: this is a case of the amazing and ubiquitous power of viruses to dramatically reshape the genetic landscape.

Cheap, powerful sequencing has enabled scientists to sample the viral world like never before. Our understanding of marine viruses, in particular, has exploded as researchers have sequenced whatever they can find in samples of seawater.

These sequencing studies have discovered "cyanophages" - viruses that infect cyanobacteria, bacteria which do much of the world's photosynthesis. These cyanophages very commonly carry photosynthesis genes, which means that opportunities for abound for a marine organism, which is normally incapable of photosynthesis, to pick the genes necessary to maintain chloroplasts, the cellular entities that enable plants, algae, and ~~cyanobacteria~~ to generate energy from sunlight.

There is, however, a high barrier to overcome before an organism can pick up genes from a bacterial virus. Typically, bacterial viruses don't infect eukaryotes, and eukaryotic viruses don't infect bacteria, due to the fact that viruses have to be niche players: they need to specialise, when it comes to manipulating cellular machinery for their own reproduction. They focus in on the unique features of their targets - in a way, they parallel computer viruses, which in most cases are OS-specific.

The most amazing example of virus-mediated gene transfer is the case of the sea slugs that produce green energy: sea slugs which eat photosynthesizing algae, and then live off photosynthesis for months. As far as I know, these are the only animals that are capable of photosynthesis, and, incredibly, there are multiple sea slug species that do this.

What happens is this: sea slugs eat algae, and, in a process named *kleptoplasty*, take up the algal chloroplasts by absorption though the gut. The sea slugs then use those stolen chloroplasts to carry out photosynthesis.

However, there's a catch: you can't maintain chloroplasts unless your own genome encodes chloroplast genes. Chloroplasts, being descendants of ancient bacterial symbionts, do have their own genomes, but their genomes do not encode all of the genes necessary for chloroplast maintenance: some contribution from the host is required. In the case of plants, algae, and ~~cyanobacteria~~ (the organisms we typically associate with photosynthesis), key chloroplast genes are found in the nuclear genome of the host.

So what about the sea slug? Animals do not carry any genes for photosynthesis. Or do they? It turns out that photosynthesizing sea slugs have managed to pick up some chloroplast genes (as discovered by sequencing portions of sea slug genomes), and these chloroplast genes show, in their surrounding DNA sequence, the hallmarks of viral transfer.

Furthermore, viral particles have been detected in both the algal chloroplasts and the cells of the sea slugs, which means that these sea slugs are constantly exposed to an opportunity for gene transfer. A few lucky gene swaps occurring over the course of a few million years of chloroplast virus exposure was apparently enough to produce sea slugs capable of photosynthesis. It's not clear what kind of viruses these are, but it's not unreasonable to suppose that they are a particular niche virus species that has gained the capability to infect both algal chloroplasts and sea slug cells.

Photosynthesizing sea slugs are just one example of the amazing gene-shuffling power of viruses. Some of the most important tumour-promoting genes (at least scientifically important, if not always clinically important), like Src, Ras, and Myc, were first discovered in viruses. These genes are not viral genes; they originated in animals, but were subsequently picked up by viruses. Since viruses shuffle in and out of genomes so frequently, it's no surprise that, over millions of years, they've had a big impact on the world's genetic landscape.

A Case for Biohydrogen

The hydrocarbon economy is faltering as oil reserves dwindle worldwide (Hirsch, 2008). Commodity prices have begun to fluctuate drastically due to the uncertain cost of petroleum, which resulted in food riots around the world in 2008. With a steadily decreasing energy supply and the demands on energy systems continually growing, the planet is in dire economic, geopolitical, and environmental straits. In order to halt the advance of climate change, prevent ecological collapse, rescue the global economy, and ensure our energy security, humanity must find a way to harness currently available (non-fossilised) energy. The largest source of energy on Earth, excluding the future potential for thermonuclear fusion reactors, is the sun. Human civilization consumes 15 TW annually while approximately 80,000 TW of solar energy fall on the Earth's surface each year (Makarieva et al., 2008). For hundreds of millions of years, this solar flux has been the driving force for life on Earth.

It is estimated that photosynthesis absorbs and distributes seven times more energy to the biosphere (~100 TW) than is consumed anthropogenically each year (Makarieva et al., 2008). There are many promising technologies in development that will cheaply harness solar output, like next generation photovoltaics, advanced wind turbines, tidal power systems, wave energy generators, solar thermal collectors, biofuels, solar fuels, electrofuels, etc. In the end, some combination of these solutions, coupled with improved energy and transportation infrastructure, will be needed to resolve our energy dilemma. In this article, I focus on cyanobacteria and their potential for hydrogen production, which, if properly developed, could provide sustainable fuel for transportation and industry in a post-petroleum world.

Cyanobacteria and the Rise of Oxygenic Photosynthesis

Approximately four billion years ago, after the formation of Earth, the first signs of life began to appear, scratching out a lithotrophic existence in an oxygen-starved atmosphere (Towe, 1996; Cleaves et al., 2008). Anoxygenic photosynthesis probably developed very early in evolution (Pierson and Olson, 1989), harnessing the power of the sun, but the real metabolic revolution didn't occur for another billion or so years, when photosynthetic eubacteria evolved the ability to split water into electrons, protons and gaseous oxygen (Nisbet et al., 2007).

This new metabolic mode, termed oxygenic photosynthesis, effectively gave living organisms access to an endless supply of electrons from water and became such a prolific process that it filled Earth's atmosphere with oxygen, fundamentally transforming the biogeochemistry of the planet (Wille et al., 2007). Moving electrons from chemically stable water molecules to high-energy carbon-carbon bonds spanned an immense reduction-oxidation (redox) gap, powered by a reliable stellar wind of solar electromagnetic radiation, which ultimately increased the size and scope of life and made the development of structurally complex eukaryotic (multicellular) organisms possible (Blankenship and Hartman, 1998). Almost all ecosystems o Earth today are directly or indirectly dependent on oxygenic photosynthesis.

The subjects of this Earth-transforming story, and the only organisms known to have evolved oxygenic photosynthesis, are the cyanobacteria (Nisbet et al., 2007). For a billion years, the electrons in water sat untouched by photosynthetic life because it was energetically unfeasible to oxidise water and reduce $NADP^+$ (an electron carrier needed for the reduction of inorganic carbon) in a single step. In an evolutionary leap, cyanobacteria pioneered the coupling of two photosynthetic reaction centres, P680 and P700 (named for the wavelengths of light they optimally absorb), referred to as photosystem II (PSII) and photosystem I (PSI), respectively (Allen and Martin, 2007; Mimuro et al., 2008). Thus, the redox problem was resolved by first pumping low-energy electrons up the redox gradient to a temporary reservoir (the quinone pool/electron transfer chain), from which the second photosystem (PSI) could pull out excited electrons in order to transfer them to an even higher energy state, sufficient for the reduction of $NADP^+$ to $NADPH + H^+$.

The rising concentration of oxygen in the atmosphere presented new challenges to early life on Earth. Most species at the time were obligate anaerobes or microaerophiles, and only those organisms that had the ability to deal with highly reactive oxygen radicals and persist in an oxygen-enriched atmosphere would inherit the majority of Earth's habitats (Brioukhanov and Netrusov, 2007; Bendall et al., 2008). Cyanobacteria, in addition to overcoming the general toxicity of oxygen (a byproduct of their new metabolism), had to find a way to acquire nitrogen under aerobic conditions (fixing atmospheric nitrogen gas into biologically-available ammonia). The molybdenum-iron nitrogenase enzyme is highly conserved among diazotrophs (nitrogen-fixers) and is very sensitive to oxygen, which destroys the activity of the enzyme. Cyanobacteria got around this issue by separating oxygenic photosynthesis and nitrogen fixation, either spatially or temporally (Tsygankov, 2007). Some cyanobacteria only activate nitrogen fixation under dark anaerobic conditions, when PSII is unable to evolve oxygen. Other cyanobacteria, like filamentous *Nostoc punctiforme*, form a specialised cell type for nitrogen fixation, called a heterocyst. Heterocysts deactivate their PSII complexes, grow thickened cell walls, and exhibit higher intracellular respiration rates, which keep oxygen levels very low (Cardona, 2009). Vegetative cells provide the heterocysts with carbohydrates, while the heterocysts provide the vegetative cells with fixed nitrogen (Cardona, 2009).

Hydrogen Metabolism in Cyanobacteria

Nitrogenase is responsible for the following reaction: N_2 + 8 H^+ + 8 $e^{"}$ + 16 ATP '! 2 NH_3 + H_2 + 16 ADP + 16 P_i (Tikhonovich and Provorov, 2007). This process is very energy-intensive, requiring 8 moles of ATP for every mole of ammonia produced. In addition to ammonia, the nitrogenase enzyme also produces a molecule of hydrogen gas for every molecule of gaseous nitrogen that it fixes. Hydrogen production via nitrogenase is relatively inefficient, due to the large amounts of ATP required, but it can still be used to produce measurable amounts molecular hydrogen.

One impediment to producing hydrogen in this way is the presence of uptake hydrogenases (Schütz et al., 2004), which oxidise nitrogenase-produced hydrogen in order to minimise the energy loss from N-fixation by reclaiming ATP via the oxyhydrogen reaction, removing oxygen from the interior of the cell and providing reducing equivalents for other cellular processes (Tamagnini et al., 2007). Concordantly, uptake-hydrogenase enzyme activity has been shown to correlate with nitrogenase activity (Schütz et al., 2004). Therefore, it is not surprising that a *Nostoc* uptake-hydrogenase knockout mutant exhibits a significantly higher hydrogen output than the wild-type (Lindberg et al., 2004).

Bidirectional hydrogenases, as their name implies, are capable of reversible catalysis of hydrogen formation or consumption ($2H^+ + 2e^- <=> H_2$). Unlike the uptake-hydrogenases, the activity of these enzymes is independent of nitrogenase activity (Schütz et al., 2004). These enzymes seem to act as putative escape valves for the excess reducing power that could build up during metabolism, but their exact function is still under debate (Tamagnini et al., 2007). Cyanobacterial bidirectional hydrogenases have a nickel-iron active site. These nickel-iron hydrogenases are more aerotolerant, but less productive than the iron hydrogenases found in other anaerobic eubacteria and in eukaryotic green algae (Ghirardi et al., 2007). Both nickel-iron and iron hydrogenases require maturation proteins in order to attain catalytic activity, but it has also been suggested that certain [Fe-Fe] hydrogenases do not require a maturation process (Asada et al., 2000; King et al., 2006). These maturation proteins are involved with the insertion of metal clustrs into the active site of the hydrogenases (Fontecilla-Camps et al., 2009). Hydrogenases are much more efficient hydrogen producers than nitrogenases, and hold great promise for future biotechnological applications (Tamagnini et al., 2007).

The overall physiology of heterocystous cyanobacteria is highly complex. Advances in genomics, transcriptomics, proteomics and metabolomics are helping scientists develop a holistic view of hydrogen metabolism (Cardona, 2009). In terms of the obstacles facing photobiological hydrogen production, they can be broken down into a few basic issues (Tamagnini et al., 2007). The first main hurdle is that the most prolific hydrogen-producing enzymes, the [Fe-Fe] hydrogenases, are highly oxygen sensitive. For a photobioreactor to be cost-effective, it should produce hydrogen under atmospheric conditions. Some investigations have looked into finding or engineering oxygen-tolerant hydrogenases (Ghirardi et al., 2007). The difficulty with oxygen-tolerant hydrogenases is that their hydrogenase activity has so far been inversely proportional to their level of oxygen tolerance ([Ni-Fe] hydrogenases).

In heterocystous cyanobacteria, this obstacle has been overcome by creating an anoxic microenvironment inside the heterocyst. Another complication arises due to the competition of different metabolic pathways for electrons from PSI. Most of the reducing power generated via oxygenic photosynthesis is diverted towards carbon and nitrogen fixation in *Nostoc punctiforme*. The goal is to better insulate energy-yielding pathways from competing metabolic processes (Agapakis et al., 2010).

Dr. Matthias Rögner, from Ruhr-Universität in Bochum, estimated at a recent conference in Sigtuna, Sweden, that under optimal growth conditions ~75% of all electrons coming from PSI could potentially be diverted to hydrogen production without negatively impacting the organism. While metabolic engineering can be vastly complex, scientists can theoretically simplify this problem by catching the electrons at their source (PSI). PSI hands its electrons off to a ferredoxin, which then shuttles the electrons away to various pathways (Tamagnini et al., 2007). One idea is to physically link a ferredoxin ligand to a hydrogenase, in order to compete with the native ferredoxins and snatch electrons away from alternate pathways and transfer them directly to the hydrogenase (Agapakis et al., 2010).

The Ideal Hydrogen-producing Organism

Nostoc punctiforme was isolated from a mutualistic association with a cycad (Costa and Lindblad, 2002). *Nostoc*, when living symbiotically, has a higher heterocyst frequency than it does as a free-living organism, and will devote much of its metabolism to the production and secretion of specific nitrogen-rich metabolites that are beneficial to its plant host (Enderlin and Meeks, 1983). Since *Nostoc* has evolved to mass-produce a specific metabolite when living symbiotically, it seems almost pre-programmed for fuel production. Heterocystous N-fixing cyanobacteria have minimal nutritional requirements, high photosynthetic efficiencies and can create anoxic microenvironments inside specialised cells that allow anaerobic processes to occur under aerobic conditions. All of these attributes are conducive to affordable bioreactor design for hydrogen production. If *Nostoc* hydrogen metabolism can be effectively engineered, using novel synthetic biology tools, the door to developing successful photobiological hydrogen-yielding technologies is opened.

Synthetic Biology: An Emerging Field

An astounding result of recent genomic sequencing projects is that the length of a genome does not predict the morphological or physiological complexity of an organism. For example, length of the human genome is similar to that of the fruit fly (Mukherji and van Oudenaarden, 2009). Instead, it has been found that biological modularity can help explain the diversity of form and function in the natural world (Mukherji and van Oudenaarden, 2009). A limited subset of predictable biological "parts" can be assembled in various ways to produce molecular "devices", which can be arranged into "systems" to carry out different functions. In this light, one can view a living cell as a combination of co-regulated genetic circuits, working in tandem. Another surprising discovery made recently shows the inherent resiliency of biological circuitry to rewiring.

Isalan et al. (2008), in order to test the limits of perturbing regulatory networks in biological systems, found that randomly rewiring *Escherichia coli* transcriptional networks by synthetically altering transcription factor/promoter pairings only resulted in faulty growth in 5% of cases. This level of tolerance to random restructuring of biological networks is highly conducive to the adaptation and evolvability of living systems by allowing large-scale alterations to be made to an organisms genome without significantly impeding its growth (Isalan et al., 2008; Mukherji and van Oudenaarden, 2009). Biological modularity and network resiliency provide powerful mechanisms for the rapid evolution of novel or optimised processes by reshuffling pre-existing genes/proteins (Isalan et al., 2008; Mukherji and van Oudenaarden, 2009; Picataggio, 2009).

As our understanding of biology grows in breadth and in depth, we come closer to the goal of being able to rationally design biological systems. Thanks to the enormous accumulation of whole-system biological data and the discovery of the modular nature of genetic and enzymatic elements during the past few decades, coupled with advances in *in silico* data analysis/modelling and rapid *in vitro* DNA synthesis technology, a new field, called synthetic biology, has emerged (Picataggio, 2009).

Synthetic biology allows for the rational manipulation of microbial phenotypes by combining systems biology, bioinformatics, protein design and engineering. With a comprehensive understanding of the molecular regulation of gene expression and protein function, biologists can begin to assemble a toolbox of reliable promoters, repressors, activators, ribosomal binding sites, reporters, signalling devices and enzymes, that can be used to design metabolic circuits in a cellular chassis - an autonomous self-replicating framework, or superstructure, that acts as the platform for synthetic circuitry (Picataggio, 2009). Standardisation of genetic tools will streamline process engineering and expand the potential to quickly develop microbial systems for the production of renewable fuels and high-value molecules (Picataggio, 2009).

Synthetic biologists have already succeeded in characterising thousands of biological parts with defined functions and performance parameters, which can be accessed by the Massachusetts Institute of Technology, but they are not yet capable of engineering whole biological systems with the same precision and reliability that, say, electrical engineers are accustomed to. One of the main challenges is to identify the subset of genes that are absolutely necessary for the survival of a minimal genome - the smallest number of genes that allows for the replication of an organism in a particular environment (Cho et al., 1999). Until researchers can build a living cell from the ground up, they won't totally understand the limits of metabolic engineering. The immense potential for engineering crucial synthetic metabolic circuits has already been demonstrated by the work of Dr. Jay Keasling, who produced an important precursor to the antimalarial drug artemisinin in *E. coli* (Hale et al., 2007). Keasling's work has cut the cost of artemisinin by ten fold and will provide many people in the third world with access to crucial malaria treatments, literally saving *millions* of lives

E. coli is an ideal cellular chassis for molecular reprocessing of low-value substrates into high-value products, but in order to use synthetic biology to tackle humanity's energy needs, it is important that we move away from heterotrophic organisms and focus on developing a photosynthetic chassis that can harness the power of the sun. Approaching solar fuel production via direct synthesis from sunlight avoids the drawbacks of traditional fermentation-based methods whose biomass feedstocks often compete directly with food crops (Tenenbaum, 2008). Researchers at UC Davis have recently developed efficient synthetically-derived fuel (isobutyraldehyde, which can be converted to isobutanol) from cyanobacteria (Atsumi et al., 2009). This result is encouraging, and suggests that the field of photosynthetic fuels will continue to grow. In the absence of a minimal genome, heterocystous cyanobacteria seem to be ideal chassis for hydrogen production.

The Development of Synthetic Biology Tools for Cyanobacteria

The creation of standard biological parts (promoters, ribosomal binding sites, repressors, activators, etc.) for cyanobacteria is simplified by the work already completed in *E. coli* and by the current open-source nature of synthetic biology resources, fostered by organisations like the Biobricks Foundation and iGEM. With our current understanding of transcriptional and translational regulation in cyanobacteria, we can redesign genetic regulatory elements and codon-optimised genes from distantly related organisms *in silico* and synthesize these constructs *in vitro* for expression in cyanobacteria. A team in France has developed codon-optimised (for expression in *E. coli, Synechocystis, Nostoc* and *Anabaena*) synthetic [Fe-Fe] hydrogenase genes from *Chlamydomonas reinhardtii* and *Clostridium acetobutylicum* that have been linked to a synthetic ferredoxin ligands derived from a chlamydomonal ferredoxin (Jaramillo, A., 2009, École Polytechnique, Palaiseau, France, privileged information). Recently, researchers in the Department of Photochemistry and Molecular Science at Uppsala University have characterised P_{trc} promoters, derived from the *lac*UV5 promoter, ribosomal binding sites and an expression vectors (pPMQAK1 and pPMQAC1) that are broadly functional in *E. coli* and in cyanobacteria (Brosius et al., 1985; Huang et al., 2010; Gibbons, 2010). With these tools, it is now possible to express prolific [Fe-Fe] hydrogenases and their respective maturation proteins in cyanobacteria. It is only a matter of time before viable hydrogen-yielding systems are in place.

Conclusion

In this post, I wanted to provide a glimpse of recent developments in photobiological hydrogen production (a field that I'm familiar with), but this is just one small piece of the puzzle. There are many teams of brilliant researchers around the globe chipping away at their own corners of sustainable energy. The future green economy will have to take a multi-faceted approach to energy, incorporating dozens of the most successful technologies, alongside behavioural and policy reform (i.e. less consumption, governmental regulation of greenhouse gasses, more localised food production, ecosystem preservation/ restoration etc.). My point was to inspire hope, despite all the doom and gloom surrounding climate change, by illuminating a subset of the work and ideas of our best and brightest, who toil day and night to build a better tomorrow. While we cannot ignore the perils that lie ahead, we must maintain a certain degree of hope in order to keep our heads above water and look towards a better horizon. Fear not, fellow humans - for every Senator James Inhofe, there are a hundred thousand scientists, social activists, community horticulturists and conservationists. By hook or by crook, we will shape our world into a more verdant and peaceful place (or die trying). We've got soul, but we're not soldiers.

Deep Ocean: Mother Nature's Very Own Science Fiction Movie

Contrary to popular belief (trekkies especially) space is not yet the final frontier - we still have plenty of unexplored frontier closer to home yet farther away from pop culture imagination: the oceans. The deep ocean remains one of the last truly explored regions, and it remains almost as mysterious as any distant galaxy.

According to Discovery Channel's epic documentary, Blue Planet, Seas of Life, "Over 60% of our planet is covered by water more than a mile deep. The deep sea is the largest habitat on earth and is largely unexplored. More people have travelled into space than have travelled to the deep ocean realm."

Our imaginations of the deep ocean verge on the alien, and as more information is uncovered, and more species discovered, we are shocked into amazement that the planet that we know so intimately, can support such alien life.

The deep ocean, including the benthic and abyssal zones makes up more than 80% of the ocean biome. These areas are cold and dark, being far below the reach of sunlight. Life is not as frequent as there are no plants or algae to feed on, and animals who call these regions home must find other means to live.

The deep ocean has markedly different physical characteristics from other parts of the Earth and even other parts of the ocean. Different pressures, temperatures, salinity, oxygen and other factors cause animals to adapt to these environments, developing unique tools of predation, diversion, and reproduction.

Light is one of the biggest factors that influence the direction of development in the deep ocean. There is no sunlight that reaches the waters of the deep ocean, and so this "twilight zone" is only lit by bioluminescent animals, which produce light using chemicals within their own bodies. Most animals need additional senses besides sight to navigate the deep ocean and survive either by catching prey or avoiding predators. Bioluminescent animals can use their light to attract prey, deflect predators, and communicate with other members of their species. This production of light can be the difference between being eaten, and escape, and also for securing your lunch.

One such predator is the deep sea angler fish. This predator uses a number of methods for effective survival in the deep ocean. Instead of wasting energy in vain pursuit of food, it sits and waits, preferring to lure and ambush its prey. Its lure is an elongated dorsal fin equipped with bioluminescent photophores at the tip of the fin. The small glowing portion of the fin draws in smaller fish in search of food, and then the fish is ambushed when within striking range of the angler's teeth. The angler fish's teeth are long, sharp and curved inward, to ensure no escape from prey. This fish more closely resembles a horror film than it does represent Mother Nature.

Ctenophores, commonly called Sea Gooseberries or Comb Jellies have a distinct pattern of eight rows (comb rows) of cilia, used for locomotion. The ciliated plates diffract light and cause a rainbow pattern of light to flash on the comb plates. Despite their mesmerizing beauty, the light attracts small prey which the Jellies consume whole.

Additionally many deep sea fish have large eyes, which help capture any residual light. This is the case with the strange *Micropinna microstoma*, which has tubular eyes shielded by a transparent head.

The mechanism for the tubular eyes was recently discovered by researchers Bruce Robinson and Kim Reisenbichler, who found that the eyes can rotate within the transparent shield. The eyes can face upward when the fish is searching for food, and forward while feeding. Macropinna is one example of the development of the "barrel eye" whose tubular shape is ultrasensitive to capturing any remaining sunlight in the water. The transparent head, allows the eyes, while facing upward, to search for shadow, throwing

Light is not the only factor affecting life in the deep ocean, however. Pressure is another important influence. The pressure underwater increases 1 atmosphere for every 10 metres of depth. Pressures in the deep ocean can reach 1,000 atm. This makes studying the deep ocean difficult, since observation is limited and collected samples have to be pressurised to maintain stability. Animals in the deep ocean have adapted to the higher pressures through bodies without excess cavities that could collapse under intense pressure. Additionally, bones and flesh material are more gelatinous and soft which helps to withstand increased pressure. As a result, there are more jellyfish-like animals, worms and soft crustaceans than bony fish residing in the deep ocean.

One most amazing mollusc is the famed giant squid. Legends of the giant squid have been around since ancients have taken to the sea, and tales of a giant creature wraps its tentacles around ships and drags vessel and crew to the ocean depths. Although unlikely that any such creature is the cause of a sailor's watery grave, tales of the giant squid do have some merit. Specimens of the giant squid have been well documented, and samples collected around the world. But an even more amazing creature exists: the colossal squid. That's right, there is a squid that is bigger than giant. The Colossal squid, *Mesonychoteuthis hamiltonia*, lives in estimated depths of 2,000-2,200 metres. In 2007, a live colossal squid was brought to the surface near New Zealand, and was measured to be 10 metres long and weighed over 1,000 pounds. This half-ton behemoth is the largest invertebrate ever captured. Scientists identified it as a male of the species, and speculate that females could be even larger. Unlike giant squid, the colossal squid have swivelling hooks on each of the sucker disks at the ends of the feeding tentacles. These tools help this colossal predator wrangle with prey, and escape predators, most notably the Antarctic sperm whale.

Food sources are rare in the deep ocean, as the lack of sunlight and oxygen prevent algae and other plants to inhabit the depths. Life, however, seems to find new sustenance nearer hydrothermal vents. Dives near the Galapagos at depths of 2,700 km led to the discovery of a new ecosystem where life is sustained by the hot, mineral rich water which erupts from geysers in the seafloor. Heated by magma, the vents are found in or near mid-ocean ridges where new sea floor is formed. Metal sulfides are expelled along with other nutrients and communities of worms, and other animals thrive in the sunless environment. Most animals have special bacteria that turn hydrogen sulfide into food.

One such creature discovered near a hydrothermal vent is the Yeti Crab. This deep sea crab is blind, and is covered in long yellow hairs, or setae, which give the crab its name. This mythical crab was discovered near Easter Island, in the vents along the Pacific-Antarctic ridge by the research submarine *Alvin*. In an attempt to explain how creatures colonise the hydrothermal vents, specimens of the crab was collected and revealed that the crab's hairy limbs support large colonies of bacteria. This supports scientist's speculations that the crab in fact, farms the bacteria which live on the hydrogen sulfide possibly as a source of food. The hairs may also act as chemical and physical sensors to help find food or mates.

Although scientists have a good idea of what might be supporting this crab and many other creatures in the deep sea, much of it is still a mystery, as is the majority of how life works in the deep ocean. Although so close to home, the deep ocean remains virtually untouched, waiting for its secrets to be discovered

Hydrogen Storage Gets New Hope

A new method for "recycling" hydrogen-containing fuel materials could open the door to economically viable hydrogen-based vehicles. In an article appearing today in Angewandte Chemie, Los Alamos National Laboratory and University of Alabama researchers working within the U.S. Department of Energy's Chemical Hydrogen Storage Centre of Excellence describe a significant advance in hydrogen storage science.

Hydrogen is in many ways an ideal fuel for transportation. It is abundant and can be used to run a fuel cell, which is much more efficient than internal combustion engines. Its use in a fuel cell also eliminates the formation of gaseous byproducts that are detrimental to the environment.

For use in transportation, a fuel ideally should be lightweight to maintain overall fuel efficiency and pack a high energy content into a small volume. Unfortunately, under normal conditions, pure hydrogen has a low energy density per unit volume, presenting technical challenges for its use in vehicles capable of travelling 300 miles or more on a single fuel tank—a benchmark target set by DOE.

Consequently, until now, the universe's lightest element has been considered by some as a lightweight in terms of being a viable transportation fuel.

In order to overcome some of the energy density issues associated with pure hydrogen, work within the Chemical Hydrogen Storage Centre of Excellence has focused on using a class of materials known as chemical hydrides. Hydrogen can be released from these materials and potentially used to run a fuel cell. These compounds can be thought of as "chemical fuel tanks" because of their hydrogen storage capacity.

Ammonia borane is an attractive example of a chemical hydride because its hydrogen storage capacity approaches a whopping 20 percent by weight. The chief drawback of ammonia borane, however, has been the lack of energy-efficient methods to reintroduce hydrogen back into the spent fuel once it has been released. In other words, until recently, after hydrogen release, ammonia borane couldn't be adequately recycled.

Los Alamos researchers have been working with University of Alabama colleagues on developing methods for the efficient recycling of ammonia borane. The research team made a breakthrough when it discovered that a specific form of dehydrogenated fuel, called polyborazylene, could be recycled with relative ease using modest energy input. This development is a significant step toward using ammonia borane as a possible energy carrier for transportation purposes.

"This research represents a breakthrough in the field of hydrogen storage and has significant practical applications," said Dr. Gene Peterson, leader of the Chemistry Division at Los Alamos. "The chemistry is new and innovative, and the research team is to be commended on this excellent achievement."

The Chemical Hydrogen Storage Centre of Excellence is one of three Centre efforts funded by DOE. The other two focus on hydrogen sorption technologies and storage in metal hydrides. The Centre of Excellence is a collaboration between Los Alamos, Pacific Northwest National Laboratory, and academic and industrial partners.

Referring to the work described in the *Angewandte Chemie* article, Los Alamos researcher John Gordon, corresponding author for the paper, stated, "Collaboration encouraged by our Centre model was responsible for this breakthrough. At the outset there were myriad potential reagents with which to attempt this chemistry."

"The predictive calculations carried out by University of Alabama professor Dave Dixon's group were crucial in guiding the experimental work of Los Alamos postdoctoral researcher Ben Davis," Gordon added. "The excellent synergy between these two groups clearly enabled this advance."

The research team currently is working with colleagues at The Dow Chemical Company, another Centre partner, to improve overall chemical efficiencies and move toward large-scale implementation of hydrogen-based fuels within the transportation sector.

Climate Sensitivity to CO2 More Limited than ExtremePprojection

A new study suggests that the rate of global warming from doubling of atmospheric carbon dioxide may be less than the most dire estimates of some previous studies and, in fact, may be less severe than projected by the Intergovernmental Panel on Climate Change report in 2007. Authors of the study, which was funded by the National Science Foundation's Paleoclimate Program and published online this week in the journal *Science*, say that global warming is real and that increases in atmospheric CO_2 will have multiple serious impacts.

However, the most Draconian projections of temperature increases from the doubling of CO_2 are unlikely.

"Many previous climate sensitivity studies have looked at the past only from 1850 through today, and not fully integrated paleoclimate date, especially on a global scale," said Andreas Schmittner, an Oregon State University researcher and lead author on the Science article. "When you reconstruct sea and land surface temperatures from the peak of the last Ice Age 21,000 years ago which is referred to as the Last Glacial Maximum and compare it with climate model simulations of that period, you get a much different picture.

"If these paleoclimatic constraints apply to the future, as predicted by our model, the results imply less probability of extreme climatic change than previously thought," Schmittner added.

Scientists have struggled for years trying to quantify "climate sensitivity" which is how Earth will respond to projected increases of atmospheric carbon dioxide. The 2007 IPCC report estimated that the air near the surface of Earth would warm on average by 2 to 4.5 degrees (Celsius) with a doubling of atmospheric CO_2 from pre-industrial standards. The mean, or "expected value" increase in the IPCC estimates was 3.0 degrees; most climate model studies use the doubling of CO_2 as a basic index.

Some previous studies have claimed the impacts could be much more severe as much as 10 degrees or higher with a doubling of CO_2 although these projections come with an acknowledged low probability. Studies based on data going back only to 1850 are affected by large uncertainties in the effects of dust and other small particles in the air that reflect sunlight and can influence clouds, known as "aerosol forcing," or by the absorption of heat by the oceans, the researchers say.

To lower the degree of uncertainty, Schmittner and his colleagues used a climate model with more data and found that there are constraints that preclude very high levels of climate sensitivity.

The researchers compiled land and ocean surface temperature reconstructions from the Last Glacial Maximum and created a global map of those temperatures. During this time, atmospheric CO_2 was about a third less than before the Industrial Revolution, and levels of methane and nitrous oxide were much lower. Because much of the northern latitudes were covered in ice and snow, sea levels were lower, the climate was drier (less precipitation), and there was more dust in the air.

All these factor, which contributed to cooling Earth's surface, were included in their climate model simulations.

The new data changed the assessment of climate models in many ways, said Schmittner, an associate professor in OSU's College of Earth, Ocean, and Atmospheric Sciences. The researchers' reconstruction of temperatures has greater spatial coverage and showed less cooling during the Ice Age than most previous studies.

High sensitivity climate models more than 6 degrees suggest that the low levels of atmospheric CO_2 during the Last Glacial Maximum would result in a "runaway effect" that would have left Earth completely ice-covered.

"Clearly, that didn't happen," Schmittner said. "Though the Earth then was covered by much more ice and snow than it is today, the ice sheets didn't extend beyond latitudes of about 40 degrees, and the tropics and subtropics were largely ice-free except at high altitudes. These high-sensitivity models overestimate cooling."

On the other hand, models with low climate sensitivity less than 1.3 degrees underestimate the cooling almost everywhere at the Last Glacial Maximum, the researchers say. The closest match, with a much lower degree of uncertainty than most other studies, suggests climate sensitivity is about 2.4 degrees.

However, uncertainty levels may be underestimated because the model simulations did not take into account uncertainties arising from how cloud changes reflect sunlight, Schmittner said.

Reconstructing sea and land surface temperatures from 21,000 years ago is a complex task involving the examination of ices cores, bore holes, fossils of marine and terrestrial organisms, seafloor sediments and other factors. Sediment cores, for example, contain different biological assemblages found in different temperature regimes and can be used to infer past temperatures based on analogs in modern ocean conditions.

"When we first looked at the paleoclimatic data, I was struck by the small cooling of the ocean," Schmittner said. "On average, the ocean was only about two degrees (Celsius) cooler than it is today, yet the planet was completely different huge ice sheets over North America and northern Europe, more sea ice and snow, different vegetation, lower sea levels and more dust in the air.

"It shows that even very small changes in the ocean's surface temperature can have an enormous impact elsewhere, particularly over land areas at mid- to high-latitudes," he added.

Schmittner said continued unabated fossil fuel use could lead to similar warming of the sea surface as reconstruction shows happened between the Last Glacial Maximum and today.

"Hence, drastic changes over land can be expected," he said. "However, our study implies that we still have time to prevent that from happening, if we make a concerted effort to change course soon."

Other authors on the study include Peter Clark and Alan Mix of OSU; Nathan Urban, Princeton University; Jeremy Shakun, Harvard University; Natalie Mahowald, Cornell University; Patrick Bartlein, University of Oregon; and Antoni Rosell-Mele, University of Barcelona.

Ancient Environment Found to Drive Marine Biodiversity

Much of our knowledge about past life has come from the fossil record but how accurately does that reflect the true history and drivers of biodiversity on Earth? "It's a question that goes back a long way to the time of Darwin, who looked at the fossil record and tried to understand what it tells us about the history of life," says Shanan Peters, an assistant professor of geoscience at the University of Wisconsin-Madison.

In fact, the fossil record can tell us a great deal, he says in a new study. In a report published on Nov. 25 in *Science* magazine, he and colleague Bjarte Hannisdal, of the University of Bergen in Norway, show that the evolution of marine life over the past 500 million years has been robustly and independently driven by both ocean chemistry and sea level changes.

The time period studied covered most of the Phanerozoic eon, which extends to the present and includes the evolution of most plant and animal life.

Hannisdal and Peters analysed fossil data from the Paleobiology Database along with paleoenvironmental proxy records and data on the rock record that link to ancient global climates, tectonic movement, continental flooding, and changes in biogeochemistry, particularly with respect to oxygen, carbon, and sulfur cycles. They used a method called information transfer that allowed them to identify causal relationships not just general associations between diversity and environmental proxy records.

"We find an interesting web of connections between these different systems that combine to drive what we see in the fossil record," Peters says. "Genus diversity carries a very direct and strong signal of the sulfur isotopic signal. Similarly, the signal from sea level, how much the continents are covered by shallow seas, independently propagates into the history of marine animal diversity."

The dramatic changes in biodiversity seen in the fossil record at many different timescales including both proliferations and mass extinctions as marine animals diversified, evolved, and moved onto land likely arose through biological responses to changes in the global carbon and sulfur cycles and sea level through geologic time.

The strength of the interactions also shows that the fossil record, despite its incompleteness and the influence of sampling, is a good representation of marine biodiversity over the past half-billion years.

"These results show that the number of species in the oceans through time has been influenced by the amount and availability of carbon, oxygen and sulfur, and by sea level," says Lisa Boush, program director in the National Science Foundation's Division of Earth

Sciences, which funded the research. "The study allows us to better understand how modern changes in the environment might affect biodiversity today and in the future." Peters says the findings also emphasize the interconnectedness of physical, chemical, and biological processes on Earth.

"Earth systems are all connected. It's important to realise that because when we perturb one thing, we're not just affecting that one thing. There are consequences throughout the whole Earth system," he says. "The challenge is understanding how perturbation of one thing for example, the carbon cycle will eventually affect the future biodiversity of the planet."

Researchers Pinpoint Date and Rate of Earth's Most Extreme Extinction

It's well known that Earth's most severe mass extinction occurred about 250 million years ago. What's not well known is the specific time when the extinctions occurred. A team of researchers from North America and China have published a paper in *Science* which explicitly provides the date and rate of extinction. "This is the first paper to provide rates of such massive extinction," says Dr. Charles Henderson, professor in the Department of Geoscience at the University of Calgary and co-author of the paper: Calibrating the end-Permian mass extinction. "Our information narrows down the possibilities of what triggered the massive extinction and any potential kill mechanism must coincide with this time."

About 95 percent of marine life and 70 percent of terrestrial life became extinct during what is known as the end-Permian, a time when continents were all one land mass called Pangea. The environment ranged from desert to lush forest. Four-limbed vertebrates were becoming diverse and among them were primitive amphibians, reptiles and a group that would, one day, include mammals.

Through the analysis of various types of dating techniques on well-preserved sedimentary sections from South China to Tibet, researchers determined that the mass extinction peaked about 252.28 million years ago and lasted less than 200,000 years, with most of the extinction lasting about 20,000 years.

"These dates are important as it will allow us to understand the physical and biological changes that took place," says Henderson. "We do not discuss modern climate change, but obviously global warming is a biodiversity concern today. The geologic record tells us that 'change' happens all the time, and from this great extinction life did recover." There is ongoing debate over whether the death of both marine and terrestrial life coincided, as well as over kill mechanisms, which may include rapid global warming, hypercapnia (a condition where there is too much CO_2 in the blood stream), continental aridity and massive wildfires. The conclusion of this study says extinctions of most marine and terrestrial life took place at the same time. And the trigger, as suggested by these researchers and others, was the massive release of CO_2 from volcanic flows known as the Siberian traps, now found in northern Russia.

Henderson's conodont research was integrated with other data to establish the study's findings. Conodonts are extinct, soft-bodied eel-like creatures with numerous tiny teeth that provide critical information on hydrocarbon deposits to global extinctions.

Princeton Release: Massive Volcanoes, Meteorite Impacts Delivered One-two Death Punch to Dinosaurs

A cosmic one-two punch of colossal volcanic eruptions and meteorite strikes likely caused the mass-extinction event at the end of the Cretaceous period that is famous for killing the dinosaurs 65 million years ago, according to two Princeton University reports that reject the prevailing theory that the extinction was caused by a single large meteorite. Princeton-led researchers found that a trail of dead plankton spanning half a million years provides a timeline that links the mass extinction to large-scale eruptions of the Deccan Traps, a primeval volcanic range in western India that was once three-times larger than France. A second Princeton-based group uncovered traces of a meteorite close to the Deccan Traps that may have been one of a series to strike Earth around the time of the mass extinction, possibly wiping out the few species that remained after thousands of years of volcanic activity.

Researchers led by Princeton Professor of Geosciences Gerta Keller report this month in the *Journal of the Geological Society of India* that marine sediments from Deccan lava flows show that the population of a plankton species widely used to gauge the fallout of prehistoric catastrophes plummeted nearly 100 percent in the thousands of years leading up to the mass extinction. This eradication occurred in sync with the largest eruption phase of the Deccan Traps the second of three when the volcanoes pumped the atmosphere full of climate-altering carbon dioxide and sulfur dioxide, the researchers report. The less severe third phase of Deccan activity kept Earth nearly uninhabitable for the next 500,000 years, the researchers report. A substantially weaker first phase occurred roughly 2.5 million years before the second-phase eruptions.

Princeton University researchers found that massive, prolonged eruptions of the Deccan Traps in India gradually eliminated species and resulted in the Cretaceous-Tertiary mass extinction that killed the dinosaurs 65 million years ago. Marine sediment trapped between Deccan lava flows revealed that a species known as *planktonic foraminifera* widely used to gauge the severity of prehistoric disasters succumbed to lava mega-flows and volcano-induced environmental stress such as acid rain and drastic climate changes. As conditions on Earth worsened, large, variedspecies (left) were eliminated. The no more than seven or eight smaller species (right) that remained dwarfed further. *(Image courtesy of Gerta Keller)*

Another group based in Keller's lab found evidence in Indian sediment of a meteorite strike from the time of the mass extinction that would have been sufficient to finish off the few but weakened species that survived the Deccan eruptions, according to a report in the journal *Earth and Planetary Science Letters* (EPSL) in October. This same sediment located in Meghalaya, India, more than 600 miles east of the Deccan Traps portrayed Earth during this period as a harsh environment of acid rain and erratic global temperatures.

Taken together, Keller said, the Princeton findings could finally put to rest the theory that the mass-extinction event known as the Cretaceous-Tertiary, or KT, for the periods it straddles was triggered solely by a large meteorite impact near Chicxulub in present-day Mexico. That impact which occurred around the time of the second-phase Deccan eruptions is thought to have been 2 million times more powerful than a hydrogen bomb and generated an enormous dust cloud and gases that radically altered the climate. Keller has long held that the Chicxulub impact was not catastrophic enough to cause the KT mass extinction the newest work from her lab, however, shows that the largest Deccan eruptions were.

"Our work in Meghalaya and the Deccan Traps provides the first one-to-one correlation between the mass extinction and Deccan volcanism," said Keller, who is lead author of the Geological Society paper and second author of the EPSL paper after lead author Brian Gertsch, who earned his Ph.D. from Princeton in 2010. Gertsch is now a postdoctoral researcher at the Massachusetts Institute of Technology.

"We demonstrate a clear cause-and-effect relationship that these massive volcanic eruptions were far more destructive than previously thought and could have caused the KT mass extinction even without the addition of large meteorite impacts," Keller said. "But given the environmental instability caused by the massive Deccan eruptions, an impact could easily have killed off the few survivor species at the end of the Cretaceous. It would have been a double whammy."

Vincent Courtillot, a geophysicist and professor at Paris University Diderot, said that the Princeton papers are based on a closer examination of Deccan volcanism and its aftermath than has been conducted previously. As such, he said, the researchers' "impressive analysis" confirms the timing of the Deccan eruptions and environmental fallout reported in recent years by various research teams, including his own.

Courtillot, who is familiar with the Princeton work but had no role in it, led the team that reported in the Journal of Geophysical Research in 2009 that Deccan volcanism occurred in three phases, the second and largest of which coincides with the Cretaceous-Tertiary mass extinction; the Keller-led study published in the Journal of the Geological Society of India confirms the second and third phases, he said.

"The significance of this recent work is that the analysis was conducted in important sections near the volcanic action, and not thousands of kilometres away as had been the case previously," Courtillot said. "They provide support for the idea that carbon and sulfur dioxide emissions were the principal agents of environmental change and stress, and conclude that the characteristics of the second-phase eruptions were such that it could alone have caused the mass extinction."

In addition, Courtillot said, the approach the teams used could prove valuable to understanding the part volcanoes played in other extinction events in Earth's history. "Exceptional, massive volcanism, I am now quite sure, is the general cause of mass extinctions," he said. "But in order to be considered as proven and quantitatively explained, we need the kind of extensive, detailed work described by these teams to be conducted for all other extinctions."

The case for Deccan over the Chicxulub impact as the Cause of the KT Extinction

Keller is prominent among scientists who reject the Chicxulub impact's role in the end-Cretaceous mass extinction. She is well known for leading a team of researchers who announced in 2003 that a sediment core from the Chicxulub crater revealed that the impact predated the mass-extinction event by about 300,000 years.

Keller and her co-authors published their findings in the journal *Proceedings of the National Academy of Sciences* in 2004 and suggested that the Chicxulub meteorite was instead one of several meteorite strikes that occurred in the several hundred thousand years leading up to the mass-extinction event.

They concluded that while destructive, the Chicxulub impact was not powerful enough to have caused widespread annihilation. Keller and her collaborators have since supported these findings with additional evidence from Texas and northeastern Mexico published in *EPSL* in 2007 and the *Journal of the Geological Society of London* in 2009, respectively.

Keller has joined other scientists in focusing her research on the 30 year old idea first championed by Virginia Tech geologist Dewey McLean that Deccan volcanism was the root of the Cretaceous mass extinction. Until recently, the theory was in question because the eruptions were thought to have been stretched out over a period of more than 1 million years, leaving plenty of time for Earth to recover between eruptions, Keller said.

Improved dating technology, however, allowed scientists particularly the team led by Courtillot to narrow the time of the largest eruptions to a few hundred thousand years at the end of the Cretaceous. Known as Deccan phase-2, this period accounted for 80 percent of the total volcanism. The first and weakest phase of activity occurred about 67.5 million years ago; the third and final eruption phase began about 300,000 years after the KT mass extinction.

In 2008, Keller and her team reported in EPSL the first direct link that the KT extinction coincided with the end of the second phase of Deccan eruptions. She explained that marine sediments preserved between lava flows from the second- and third-phase eruptions contained evidence of the KT boundary, a thin, worldwide geological layer that marks the mass-extinction event.

Deccan Volcanism Behind the Mass Extinction, So Say the Plankton

The work published Nov. 1 by the Geological Society of India builds on Keller's 2008 paper in EPSL. She and her co-authors examined cores from Deccan lava flows near Rajahmundry in the Krishna-Godavari Basin, the remnant of an ancient sea on the Bay of Bengal coast, and found that lava flows from the second and third Deccan phases are separated by marine sediments.

Keller worked with P.K. Bhowmick, H. Upadhyay, A. Dave, A.N. Reddy and B.C. Jaiprakash, scientists with India's government-operated Oil and Natural Gas Corporation, which owns the sediment cores. Also included is Thierry Adatte, a geologist with the University of Lausanne in Switzerland, who is Keller's long-time collaborator and a co-author on the papers challenging the time of the Chicxulub impact, as well as previous papers on Deccan volcanism.

The team examined the basin's sediment layers to determine the size and number of a species known as planktonic foraminifera that remained following each eruption phase. These plankton are single-celled micro-organisms ranging in size from the point of a needle to a pinhead that are highly sensitive to changes in oxygen, salinity, temperature and nutrients, Keller said. Their sensitivity to environmental changes and their near extinction at the end of the Cretaceous makes the species key to determining the timespan, pace and severity of the mass extinction.

After studying microplankton remains in sediment from below, between and above the second-phase lava flows, the researchers observed that the number of living species dropped 50 percent at the onset of eruptions. The species count plunged by another 50 percent after the first of what would be four lava mega-flows. No more than seven to eight of the species that were most tolerant to environmental changes survived after the first mega-flow, and no recovery occurred between subsequent mega-flows. By the end of the fourth mega-flow the mass extinction was complete, the researchers wrote.

The vast amounts of carbon dioxide and sulfur dioxide poured into the atmosphere by the end of the second volcanic phase estimated to be 30 times more than the levels produced by the Chicxulub impact resulted in, among other crises, heavy acid rain, acidic oceans and global temperatures that swung between scorching and frigid, the researchers report. The third eruption phase prolonged these conditions.

Thus, the number of species evolving remained low, and existing species dwarfed during the 500,000 year period after the mass extinction, although no significant extinctions occurred again, Keller and her co-authors found. New, larger marine species did not appear until after the third phase when Deccan eruptions went dormant, suggesting that life began to recover as the atmosphere became less poisonous.

"In my work, I had always observed evidence of marked changes in species abundance with gradually higher levels of stress and extinction during the last several hundred thousand years, rather than one single instantaneous annihilation," Keller said. "For lack of better evidence, scientists had interpreted this steady decline as the result of climate and sea-level changes."

Evidence that a Large Meteorite Helped Finish the Job

For the paper published Oct. 15 in EPSL, Keller and her co-authors provide a supporting and more nuanced depiction of conditions during the Deccan period. They examined sediments from an ancient shallow sea in Meghalaya where rock layers are known to contain among the clearest fossil records of the Cretaceous-Tertiary mass extinction, Keller said.

She worked with lead author Gertsch; the geologist Adatte; Rahul Garg and Vandana Prasad from the Birbal Sahni Institute of Palaeobotany in India; Zolt Berner from the Karlsruhe Institute of Technology in Germany; and Dominik Fleitmann at the University of Bern in Switzerland.

Analysis of the Meghalaya sediment revealed an inhospitable planet rife with high humidity, severe storms and massive blooms of the plankton species Guembelitria cretacea, a disaster opportunist that flourished in devastated environments when few other species survived.

At the same time, the team detected large amounts of iridium, an element typically associated with meteorite impacts, Keller said. Iridium is rare on Earth yet is found in high concentrations in the KT boundary, a phenomenon known as the iridium anomaly. Remnants of iridium at the KT boundary in Meghalaya coincide with the global KT boundary iridium anomaly, she said.

The new evidence of a meteorite strike at Meghalaya that coincides with the KT mass extinction supports the theory Keller proffered in 2003 that multiple meteorites struck Earth around the time of the Deccan eruptions, adding to the volcano-fuelled misery of the mass-extinction era.

"Our data suggest that the mass extinction of the dinosaurs and other species was caused by the harsh conditions resulting from massive Deccan eruptions and the coincidence of multiple meteorites," Keller said. "In light of this new evidence, the single-impact story seems more like an article of faith at this point."

The study published in the *Journal of the Geological Society of India* about the Deccan eruption and the meteorite research published in EPSL were both supported by grants from the National Science Foundation

California Making Headway in Battle Against Childhood Obesity but Successes are Uneven

A new study offers hope that California may finally be getting a handle on its 30 year battle with childhood obesity, but it also showcases a patchwork of progress that leaves the majority of the state's counties still registering increases in obesity rates among school-age children. According to the study, "A Patchwork of Progress: Changes in Overweight and Obesity Among California 5th, 7th and 9th Graders, 2005-2010," prepared by the UCLA Centre for Health Policy Research and the California Centre for Public Health Advocacy (CCPHA), the percentage of overweight and obese children in the state dropped 1.1 percent from 2005 to 2010. However, 38 percent of children are still affected a rate nearly three times higher than it was 30 years ago, when the obesity epidemic began.

Even more concerning, according to the lead author of the study, UCLA's Susan Babey, Ph.D., is that improvements are not being seen statewide. "Children's health is still at risk in a significant number of counties," Babey said. "We found that 31 of California's 58 counties experienced an increase in childhood overweight over the five-year period from 2005 to 2010. We hope this county-by-county analysis will help community leaders pinpoint and take action in counties in the greatest danger."

The highest rates in the state were found in Imperial (46.9 percent), Colusa (45.7 percent), Del Norte (45.2 percent) and Monterey (44.6 percent) counties. Two of those counties, Del Norte and Colusa, also had the dubious distinction of having the highest increases over the last five years (16.2 percent and 13.3 percent, respectively).

Marin County, with 24.9 percent of children overweight or obese, had the lowest level in the state. However, the Marin County rate, historically the lowest in the state, has grown 5.5 percent since 2005.

The study describes both the health and economic repercussions of elevated obesity rates. According to the study, children who are overweight or obese often grow up to be obese adults with increased risk for chronic diseases like diabetes, cardiovascular disease, strokes and some cancers. What's more, the study says, California spends more public and private money on the health consequences of obesity than any other state more than $21 billion annually.

In 2004, California began implementing a series of state laws banning sugary drinks and junk food from public school campuses. That, along with other local and statewide policies addressing the availability, marketing and promotion of unhealthy foods and an increased emphasis on healthier food and expanding opportunities for physical activity, may be contributing to the statewide improvements revealed in this study.

"California led the nation in establishing many of the most innovative programs and policies that are improving our children's chances for a healthier life," said the CCPHA's Harold Goldstein, Dr.P.H. "Increased awareness and a growing array of school and community policies and programs are beginning to have an impact. But in light of the huge number of counties where childhood obesity rates continue to climb, our efforts must continue and even expand, especially in those areas where we now know children are most at risk."

Data for the study was drawn from the California Physical Fitness Test (Fitnessgram), which is administered annually to all California public school students in grades five, seven and nine. Measured height and weight data from the test were used to calculate body mass index (BMI), and BMI was used to determine rates of overweight and obesity, based on the 2000 Centres for Disease Control and Prevention sex-specific BMI-for-age growth charts.

University of Saskatchewan, Royal Ontario Museum Researchers Track Half-billion Year Old Predator

Researchers from the University of Saskatchewan and Royal Ontario Museum (ROM) have followed fossilised footprints to a multi-legged predator that ruled the seas of the Cambrian period about half a billion years ago. "Short of finding an animal at the end of its trackway, it's really very rare to be able to identify the producer so confidently," said Nicholas Minter, lead author of the article on the study, which appears in the latest issue of the *Proceedings of the Royal Society B*. Minter is a postdoctoral research fellow in the U of S department of geological sciences.

The research team worked with samples gathered from the Burgess Shale, famed for its exquisitely detailed fossils from the Cambrian Explosion, a time when life underwent a dramatic change with the appearance of all the modern groups of organisms and some bizarre creatures. Located near the village of Field in Canada's Yoho National

Park in British Columbia, the Burgess Shale is an international treasure, providing an unparalleled window into the distant past.

Fossils from the Burgess Shale record not only the animals themselves exceedingly rare because most of them had soft bodies but also the trackways they left behind while hunting on the sea floor.

"Most researchers have focused on the body fossils of the Burgess Shale," said study co-author Gabriela Mángano, who co-leads the ichnology research group in the U of S geological sciences department with colleague Luis Buatois. "By studying its trackways, trails and burrows, we may dramatically impact our understanding of these ancient ecosystems."

Key to the research were trackways collected during a field expedition in 2008 led by ROM curator Jean-Bernard Caron.

"I spotted a portion of the largest trackway, which is over three metres in length, in 2000," he said. "At that time we left most of it behind us. We could not carry the pieces safely down slope from this remote site without helicopter support."

The 2008 expedition included this support, so fragments of the trackway were collected by carefully separating them from the associated rock layers. These delicate pieces were then packed, air lifted from the mountain, and shipped to the ROM.

Fossil trackways and other fossilised evidence of animal activities such as burrows, bite marks and feces are known as trace fossils. These provide evidence of where animals were living and what they were doing, but the full identity of the producers is rarely known.

In this case, size of the tracks and the number of legs needed to make them left only one suspect: *Tegopelte gigas*. This caterpillar-like animal sported a smooth, soft shell on its back and 33 pairs of legs beneath. One of the largest arthropods of its time, it could reach up to 30 cm in length.

By analysing both the fossilised remains of *Tegopelte* and the trackways, the researchers were able to reconstruct how this animal would have moved. The creature was capable of skimming rapidly across the seafloor, with legs touching the sediment only briefly, supporting the view that *Tegopelte* was a large and active top carnivore. Such lifestyles would have been important in shaping early marine communities and evolution during the Cambrian explosion.

The trackways were collected under Parks Canada Research and Collecting permits and are now located at the ROM. Managed by

Parks Canada, the Burgess Shale was recognised in 1981 as one ofCanada's first UNESCO World Heritage Sites. Now protected under the larger Rocky Mountain Parks UNESCO World Heritage Site, the Burgess Shale attracts visitors to Yoho each year for guided hikes to the restricted fossil beds from July to September.

The full article, "Skimming the surface with Burgess Shale arthropod locomotion," is published in the *Proceedings of the Royal Society B.* Photos and illustrations related to the article are available from the authors. Funding for this research was provided through grants from the Natural Sciences and Engineering Research Council, the Canadian Commonwealth Scholarship Program, the U of S and ROM.

Land Animals, Ecosystems Walloped after Permian Dieoff

The cataclysmic events that marked the end of the Permian Period some 252 million years ago were a watershed moment in the history of life on Earth. As much as 90 percent of ocean organisms were extinguished, ushering in a new order of marine species, some of which we still see today. But while land dwellers certainly sustained major losses, the extent of extinction and the reshuffling afterward were less clear. In a paper published in the journal *Proceedings of the Royal Society B,* researchers at Brown University and the University of Utah undertook an exhaustive specimen-by-specimen analysis to confirm that land-based vertebrates suffered catastrophic losses as the Permian drew to a close. From the ashes, the survivors, a handful of genera labelled "disaster taxa," were free to roam more or less unimpeded, with few competitors in their respective ecological niches. This lack of competition, the researchers write, caused vicious boom-and-bust cycles in the ecosystems, as external forces wreaked magnified havoc on the tenuous links in the food web. As a result, the scientists conclude from the fossil record that terrestrial ecosystems took up to 8 million years to rebound fully from the mass extinction through incremental evolution and speciation.

"It means the (terrestrial ecosystems) were more subject to greater risk of collapse because there were fewer links" in the food web, said Jessica Whiteside, assistant professor of geological sciences at Brown and co-author on the paper.

The boom-and-bust cycles that marked land-based ecosystems' erratic rebound were like "mini-extinction events and recoveries," said Randall Irmis, a co-author on the paper, who is a curator of paleontology at the Natural History Museum of Utah and an assistant professor of geology and geophysics at Utah.

The hypothesis, in essence, places ecosystems' recovery post-Permian squarely on the repopulation and diversification of species, rather than on an outside event, such as a smoothing out of climate. The analysis mirrors the conclusions reached by Whiteside in a paper published last year in *Geology*, in which she and a colleague argued that it took up to 10 million years after the end-Permian mass extinction for enough species to repopulate the ocean restoring the food web for the marine ecosystem to stabilise.

"It really is the same pattern" with land-based ecosystems as marine environments, Whiteside said. The same seems to hold true for plants, she added.

Some studies have argued that continued volcanism following the end-Permian extinction kept ecosystems' recovery at bay, but Whiteside and Irmis say there's no physical evidence of such activity.

The researchers examined nearly 8,600 specimens, from near the end of the Permian to the middle Triassic, roughly 260 million to 242 million years ago. The fossils came from sites in the southern Ural Mountains of Russia and from the Karoo Basin in South Africa. The specimen count and analysis indicated that approximately 78 percent of land-based vertebrate genera perished in the end-Permian mass extinction. Out of the rubble emerged just a few species, the disaster taxa. One of these was *Lystrosaurus*, a dicynodont synapsid (related to mammals) about the size of a German shepherd. This creature barely registered during the Permian but dominated the ecosystem following the end-Permian extinction, the fossil record showed. Why *Lystrosaurus* survived the cataclysm when most others did not is a mystery, perhaps a combination of luck and not being picky about what it ate or where it lived. Similarly, a reptilian taxon, procolophonids, were mostly absent leading to the end-Permian extinction, yet exploded onto the scene afterward.

"Comparison with previous food-web modelling studies suggests this low diversity and prevalence of just a few taxa meant that links in the food web were few, causing instability in the ecosystem and making it susceptible to boom-bust cycles and further extinction," Whiteside said.

The ecosystems that emerged from the extinction had such low animal diversity that it was especially vulnerable to crashes spawned by environmental and other changes, the authors write. Only after species richness and evenness had been re-established, restoring enough population numbers and redundanc to the food web, did the terrestrial ecosystem fully recover. At that point, the carbon cycle, a broad indicator of life and death as well as the effect of outside influences, stabilised, the researchers note, using data from previous studies of carbon isotopes spanning the Permian and Triassic periods.

"These results are consistent with the idea that the fluctuating carbon cycle reflects the unstable ecosystems in the aftermath of the extinction event," Whiteside said.

The National Science Foundation and the University of Utah funded the work. Reporters and the general public have free access to the manuscript through an award from the University of Utah J. Willard Marriott Library Open Access Publishing Fund.

New Technique Unlocks Secrets of Ancient Ocean

Earth's largest mass extinction event, the end-Permian mass extinction, occurred some 252 million years ago. An estimated 90 percent of Earth's marine life was eradicated. To better understand the cause of this "mother of all mass extinctions," researchers from Arizona State University and the University of Cincinnati used a new geochemical technique. The team measured uranium isotopes in ancient carbonate rocks and found that a large, rapid shift in the chemistry of the world's ancient oceans occurred around the extinction event. The mechanism of the end-Permian mass extinction has been much debated. One proposed cause for the extinction, the release of toxic hydrogen sulfide gas, is directly related to oceanic anoxia, which is a depletion of dissolved oxygen from the ocean.

Widespread evidence exists for oceanic anoxia before the extinction, but the timing and extent of anoxia remain unknown. Previous hypotheses posited that the deep ocean was depleted of oxygen for millions of years before the end-Permian extinction. The new research using measurements of uranium isotopes in ancient carbonate rocks indicates that the period of ocean-wide anoxia was much shorter.

"Our study shows that the ocean was anoxic for at most tens of thousands of years before the extinction event. That's much shorter than prior estimates," says Gregory Brennecka, the lead author of the study and a graduate student in ASU's School of Earth and Space Exploration in the College of Liberal Arts and Sciences.

Brennecka, working in Professor Ariel Anbar's research group, conducted the analysis of the samples. Anbar is a professor in ASU's School of Earth and Space Exploration and the Department of Chemistry and Biochemistry. Achim Herrmann, a senior lecturer at

Barrett, the Honours College at ASU, and Thomas Algeo of the University of Cincinnati, who collected the samples in China, helped guide the selection of samples and interpretation of data.

The team studied samples of carbonate rock from Dawen in southern China for uranium isotope ratios (238U/235U) and thorium to uranium ratios (Th/U). The study presumes that carbonate rocks capture 238U/235U and Th/U of the seawater in which they were deposited. If so, they can be used to study changes in the chemistry of ancient oceans. In separate, related work, the team is testing the limits of this assumption.

In a section of rock spanning the time of the extinction, the team found a marked shift in 238U/235U in the carbonate rocks immediately prior to the mass extinction, which signals an increase in oceanic anoxia. The team also found higher Th/U ratios in the same interval, which indicate a decrease in the uranium content of seawater. Lower concentrations of uranium in seawater also serve as signals of oceanic anoxia.

These decreases in 238U/235U and increases in Th/U only occur at the section of rock that contains the end-Permian extinction horizon. This shows that a period of oceanic anoxia existed only briefly prior to the mass extinction, rather than the previously hypothesized much longer timeframe.

The team's findings represent an increase in knowledge about the ocean's chemistry at a critical period of Earth's history. "This technique gives us a better understanding of how ocean chemistry can change over time, and how sensitive it is to certain environmental factors," says Brennecka.

The implications of the new geochemical tool the researchers developed are just as important as the study's findings.

Uranium isotope ratios have been utilised to study the ocean's chemistry before, but only in black shale, a different and less common type of rock. This study represents the first time uranium isotope ratios have been studied in carbonates for paleo-redox purposes, which is a promising new geochemical tool for future research.

"One of the important outcomes of this study is that we were able to quantify the relative change in the amount of oceanic anoxia across the extinction event in the global ocean. Previous studies were only able to show whether anoxic conditions existed or not.

We can now compare this event to other events in Earth history and develop a better understanding of how the amount of oxygen in the Earth's ocean has changed through time and how this might have affected marine diversity," says Herrmann.

Carbonates are much more widespread than black shales on Earth through space and time. "By focusing on carbonates we can study ancient anoxic events in many more places and times," says Anbar. "This was our major motivation in developing the uranium isotope technique."

It is only recently that researchers have developed the ability to precisely measure slight variations in uranium ratios, largely due to research completed at ASU. Most of the team's research in this study was conducted at ASU. The study samples were analysed at ASU's W. M. Keck Foundation Laboratory for Environmental Biogeochemistry.

"Over the past decade, my research group has worked with many collaborators to develop new techniques to study changes in oxygen in the Earth's ocean through time," says Anbar. "We are especially interested in the connections between ocean oxygenation and biological evolution. The uranium isotope technique is the newest method. We expect it will be very useful. This study shows that it is yielding insights pretty quickly."

"It is exciting to be here, because most of the development work to measure uranium isotopes was done at ASU over the past five years. It is exciting to be at the forefront of these advancements," says Brennecka.

Astrophysics and Extinctions: News about Planet-threatening Events

Space is a violent place. If a star explodes or black holes collide anywhere in our part of the Milky Way, they'd give off colossal blasts of lethal gamma-rays, X-rays and cosmic rays and it's perfectly reasonable to expect Earth to be bathed in them. A new study of such events has yielded some new information about the potentil effects of what are called "short-hard" interstellar radiation events. Several studies in the past have demonstrated how longer high-energy radiation bursts, such as those caused by supernovae, and extreme solar flares can deplete stratospheric ozone, allowing the most powerful and damaging forms of ultraviolet radiation to penetrate to Earth's surface.

The probability of an event intense enough to disrupt life on the land or in the oceans becomes large, if considered on geological timescales. So getting a handle on the rates and intensities of such events is important for efforts to connect them to extinctions in the fossil record.

"We find that a kind of gamma ray burst a short gamma ray burst is probably more significant than a longer gamma ray burst," said astrophysicist Brian Thomas of Washburn University.

Improved and accumulated data collected by the SWIFT satellite, which catches gamma ray bursts in action in other galaxies, is providing a better case for the power and threat of the short bursts to life on Earth.

The shorter bursts are really short: less than one second long. They are thought to be caused by the collision of two neutron stars or maybe even colliding black holes. No one is certain which. What is clear is that they are incredibly powerful events.

"The duration is not as important as the amount of radiation," said Thomas. If such a burst were to happen inside the Milky Way, it its effects would be much longer lasting to Earth's surface and oceans.

"What I focused on was the longer term effects," said Thomas. The first effect is to deplete the ozone layer by knocking free oxygen and nitrogen atoms so they can recombine into ozone-destroying nitrous oxides. These long-lived molecules keep destroying ozone until they rain out. "So we see a big impact on the ozone layer."

Those effects are likely to have been devastating for many forms of life on the surface including terrestrial and marine plants which are the foundation of the food web.

Based on what is seen among other galaxies, these short bursts, it seems that they occur in any given galaxy at a rate of about once per 100 million years. If that is correct, then it's very likely that Earth has been exposed to such events scores of times over its history. The question is whether they left a calling card in the sky or Earth's geological record.

Astronomical evidence is not likely, said Thomas, because the galaxy spins and mixes pretty thoroughly every million years, so any remnants of blasts are probably long gone from view. There might, however, be evidence in the ground here on Earth, he said. Some researchers are looking at the isotope iron-60, for instance, which has been argued as a possible proxy for radiation events. If isotopes like iron-60 can reveal the strata of the events, it then becomes a matter of looking for extinction events that correlate and seeing what died and what survived which could shed more light on the event itself.

"I work with some paleontologists and we try to look for correlations with extinctions, but they are skeptical," said Thomas. "So if you go and give a talk to paleontologists, they are not quite into it. But to astrophysicists, it seems pretty plausible."

Thomas will be presenting his work on October 9, 2011, at the annual meeting of the Geological Society of America in Minneapolis. This work was supported by the NASA Astrobiology: Exobiology and Evolutionary Biology Program.

6

Marine Life in the Antarctic

The Southern Ocean covers 35 million km ^{2} and comprises about 10% of the Earth's oceans. Of the 4.6 million km ^{2} of continental shelf, one-third is covered by floating ice shelves (Clarke & Johnston 2003). The sea ice oscillates between a coverage of 60% in winter and 20% in summer and is, together with the sea beneath, the main driver of the Antarctic ecosystem and the Earth's ocean circulation. These conditions have caused a partial isolation of the ecosystem in the past 30 million years, and the unique environment has allowed an evolutionary dispersal of Antarctic species into the adjacent ocean's deep sea and vice versa. Recent ecological conditions in Antarctic waters not only attract the charismatic great whales, but also birds and deep-sea invertebrates from the entire world's ocean.

The Census of Marine Life recognised that the Southern Ocean is home of a key component of the Earth's biosphere and launched the Census of Antarctic Marine Life (CAML) in 2005, considered the major marine biodiversity contribution to the International Polar Year 2007–8. It followed international initiatives such as the SCAR projects "BIOMASS", "Ecology in the Antarctic Sea-Ice Zone (EASIZ, Arntz & Clarke 2002; Clarke et al. 2006), "Evolution in the Antarctic" (EVOLANTA, Eastman et al. 2004), and "Evolution and Biodiversity in the Antarctic" (EBA, results of the 10th SCAR-Biology Symposium to be published as a special volume of Polar Science) as well as the projects "European Polarstern Study" (EPOS, Hempel 1993), "Investigación Biológica Marina en Magallanes relacionada con la Antártida" (IBMANT, Arntz & Ríos 1999; Arntz et al. 2005), "Antarctic benthic DEEP-sea biodiversity (ANDEEP, Brandt & Ebbe 2007), and "Latitudinal Gradient Project" (LGP, Balks et al. 2006).

Consequently, CAML was based on a very active international scientific community and covered a broad spectrum of organisms ranging from microbes to mammals. It cooperated closely with other Census projects, especially the Ocean Biogeographic Information System (OBIS), Census of Marine Zooplankton (CMarZ), Biogeography of Deep-Water Chemosynthetic Ecosystems (ChEss), Arctic Ocean Diversity (ArcOD), and Census of Diversity of Abyssal Marine Life (CeDAMar), because of two aspects. First, by combining all three other oceans by the Antarctic Circumpolar Current (ACC), the Southern Ocean provides a link for most large marine ecosystems. Second, a considerable part of the rich Antarctic fauna is unique and thus contributes significantly to the world's total marine biodiversity.

The scientific aim of CAML was to provide essential knowledge to answer the most challenging question of the future of the Antarctic ecosystem in a changing world. The strategic objective was to create a network of knowledge within the research community and to provide a forum for communication, including the most intensive outreach activities that ever concerned the work of Antarctic marine biologists. Thanks to the CAML two overarching initiatives, the biogeographic data portal SCAR-MarBIN and the barcoding initiative, intensified their efficiency, providing essential tools for scientists to share data. CAML was one of the leading Antarctic projects of the International Polar Year 2007–8 and was part of the biology program of the Scientific Committee on Antarctic Research (SCAR). Although the Census/CAML was able to support scientific coordination, the field work was funded by the national Antarctic research programs. This review is compiled at an early stage of CAML's synthesis phase. It provides a preliminary overview and concentrates mainly on results from core projects presented in the Genoa workshop in May 2009, to be published in Deep-Sea Research II in 2010 and edited by S. Schiaparelli et al. All references cited herein as "submitted" refer to this special CAML volume.

Environmental Settings

The extreme seasonality in the Antarctic results in a permanently dark winter and a summer with 24 hours sunshine south of 66° 33¢ S. The low temperature, and consequently the formation of the sea ice, is due to the low angle of irradiation of the sun, the high albedo of ice, and the zonal atmospheric and oceanographic circulation. The marine habitat is geographically limited to the south by a glaciated coast. The ACC combines all three other ocean basins and in the north it adjoins warmer waters at the Antarctic Convergence. Over evolutionary time the Antarctic ecosystem experienced a permanent advance and retreat of continental glaciation which started with the formation of the ACC 25 million to 30 million years ago and has continued with obvious glacial–interglacial cycles in the past 900,000 years.

History of Antarctic Research and Exploitation

The era of early naturalists was related to both the discovery of the unknown region and the exploitation of natural resources. One example is the German naturalist Georg Forster, who participated with his father Johann Reinhold in James Cook's second trip around the world (1772–75). Another example is the Weddell seal, which was named after the Scottish sealer James Weddell who in 1823 reached 74°34¢ S, the most southerly position ever reached at that time. The famous Adélie penguin was named after the wife of the French explorer Jules Dumont d'Urville, who travelled twice to Antarctica between 1838 and 1840. Milestones of taxonomic surveys (Dater 1975) started with the famous Challenger expedition (1872–76) which resulted in 38 volumes of scientific results: 4,714 new species were discovered of which several were from the Antarctic.

The Belgica undertook the first truly scientific expedition to high-latitude Antarctic waters, during which she advanced farther south than any ship before and overwintered in 1898–99 west of the Antarctic Peninsula. The Valdivia expedition of 1898–99 contributed substantially to the understanding of global oceanography and included biological deep-sea sampling in the sub-Antarctic. Highly efficient were also the German Antarctic expedition with the Gauss (1901–03), the Swedish South Polar Expedition with the Antarctica (1901–04), and the British Scotia expedition (1902–04) which conducted trawling and dredging studies of pelagic and benthic organisms. The period 1925–39 was dominated by the Discovery expeditions from which publications, including those of recent surveys, are still ongoing.

The exploitation of natural resources started at the beginning of nineteenth century. Populations of Antarctic fur and elephant seals crashed close to extinction by the 1820s. Whaling started at the beginning of the twentieth century. The biomass of the largest species – blue, fin, humpback, southern right, and sei whales – were reduced to between 50% and 0.5% of their original worldwide stock whereas the smallest, the Antarctic minke, became most abundant (Laws 1977). Thus, the natural dominance pattern of whale species was turned upside-down. Interesting calculations have been made about the negative impact of the whaling to deep-sea animals since whale carcasses have no longer been important food sources for marine organisms (Jelmert & Oppen-Berntsen 1995). Bottom trawling in the 1960s reduced the stocks of the marbled rock cod (*Notothenia rossii*) and mackerel ice fish (*Champsocephalus gunnari*) west of the Antarctic Peninsula (Kock 1992) within very short periods, and devastated slow-growing benthic communities. The exploitation of natural resources was the most effective anthropogenic impact that Southern Ocean biodiversity ever experienced. However, the hitherto inviolacy of most high-latitude Antarctic marine habitats is almost unique on Earth, but the ecosystem is increasingly threatened by the new longline fishing and by the impact of climate change

Modern Pre-CAML Biodiversity Studies

In the 1980s, ecological analyses using bulk parameters (for example, tried to solve so-called "process orientated" questions without spending much time determining species diversity. Among the few studies with high taxonomic resolution, outstanding progress was made by the work on the evolutionary radiation of fish (Eastman & Grande 1989). In this phase the macrobenthos became known to be regionally dominated by sessile suspension feeders (Bullivant 1967); their communities later turned out to be more dynamic than previously expected (Dayton 1990; Arntz & Gallardo 1994; Gutt 2000, 2006; Gutt & Piepenburg 2003; Potthoff et al. 2006; Barnes & Conlan 2007; Seiler & Gutt 2007; Smale et al. 2008). Plankton studies added substantial information to the traditional view of the simple Antarctic pelagic system consisting only of algae, krill (*Euphausia superba*), and few apex predators. Small organisms became known to contribute to the microbial loop by being relevant for the re-mineralisation in a partly iron-limited "high nutrient – low chlorophyll" system. Improved sea-ice research elucidated the diversity not only of unicellular algae but also of metazoans living in and associated with this unique habitat (Thomas & Dieckmann 2009), including the trophic key species of the Antarctic food web, the Antarctic krill (Thomas et al. 2008).

The Scientific Strategy

At first the term "census" had to been interpreted literally: species and specimens were identified and counted. Secondly, CAML researchers raised the question why some of these species co-exist in specific communities whereas others do not, the answers demanding both evolutionary and ecologically approaches at various spatial scales.

What Were the Major Gaps?

The scientific effort during the pre-CAML phase reflected the good accessibility of the area around the Antarctic Peninsula and historical developments in poorly accessible areas, for example the inner Weddell and Ross Seas with large gaps in between. The Antarctic deep sea was only known from studies with selective samples with a reduced taxonomic scope. Life in some typical Antarctic habitats was very poorly known, especially from under the ice shelves and the permanent pack ice. The biodiversity not only of microorganisms, but also of rare charismatic species, for example toothed whales, had almost been overlooked and some historic data were hardly accessible. The identification of many invertebrate eggs and larvae to the species level was impossible, and only hypotheses existed in relation to cryptic species. The question about the relation between ecosystem functioning and biodiversity has a long tradition but it is still – at least for the Antarctic – difficult to address. Finally, the pre-CAML era was characterised by the knowledge that climate change would not stop at the Antarctic Circle, but background information and observations on its impact to the ecosystem were scarce

Approaches to Closing Gaps

Core strands of CAML were scientific expeditions and the data management allowing overarching analyses. Success has also been reached through the standardisation of field methods, for example by using the standard nets, continuous plankton recorders, video-equipped remotely operated vehicles (ROVs), or sleds. The major tool for ensuring information management is the "Marine Biodiversity Information Network of SCAR" being the local node of the Census of Marine Life/ UNESCO OBIS network. It was initiated by the Royal Belgian Institute of Natural Sciences and CAML became its major research partner. So far, over 1 million georeferenced records from 156 datasets are available. The Register of Antarctic Marine Species (RAMS) comprises 6,551 primarily benthic and 702 pelagic species (as at May 2010) and is constantly updated by over 70 editors and contributing scientists (De Broyer & Danis submitted).

Datasets range from historic information going back to 1781 to recent and genetic data. A barcode manager supported CAML scientists in analysing over 11,000 sequences (Grant & Linse 2009). Thus, CAML contributes to the Barcode of Life project) and the "Fish Barcode of Life Initiative" (FISH-BOL, Spatially explicit ecological models were developed, for example to predict potential fish habitats and to simulate the succession of biodiversity after disturbance (Potthoff et al. 2006). A new tool, "GeoPhyloBuilder" and network analyses (Raymond & Hosie 2009) are being used to visualise phylogeographic data.

Evolutionary Large-Scale Patterns and Non-Circumpolar Cryptic Species

The question of bipolar species experienced a renaissance under CAML. No doubt exists about the annual pole-to-pole migrations of the blue, humpback and fin whales as well as seabirds such as the Arctic tern. In addition, a bipolar occurrence of a few benthic and pelagic invertebrate species had been controversially discussed. A recent comparison between the Register of Antarctic Marine Species and the ArcOD database revealed approximately 230 species names to which occurrences from both polar regions were attributed. Recent attempts to provide evidence for their existence with genetic methods were successful, for example for the amphipod *Eurythenes gryllus* occurring at the upper slopes of the Canadian Arctic, around Antarctica, and in the deep sea in between (France & Kocher 1996;s De Broyer et al. 2007).

Such evidence failed for the pteropod *Limacina helicina*, being so far considered as one species but having 32% divergence between both polar regions (J.M. Strugnell, unpublished observations). Another weak example is the sponge *Stylocordyla borealis*, which has two sympatric distinct growth forms even within the Antarctic, one with a thick stalk, the other like a lollipop. For the widespread and well-known deep-sea holothurian *Elpidia glacialis*, which has strong polar emergence, six subspecies are known and, using traditional methods, it is only a matter of interpretation not to consider these as six true species. A morphologic and genetic documentation of the existence of bipolar species among deep-sea komokiaceans and other foraminiferan-like protists was highlighted by Brandt et al. (2007a). A high genetic and morphologic similarity was found for the planktonic anthomedusa genus *Pandea* between the north Pacific near Japan and East Antarctica (D. Lindsay et al., unpublished observations). In conclusion, it remains open whether genetic methods will continue to confirm the bipolar occurrence of species and, consequently, gene-flow over extremely long distances or whether true bipolar species will remain rare exceptions.

Before we can understand the role of the Southern Ocean within global biodiversity patterns and underlying evolutionary processes, our knowledge of geographic coverage has to be completed, especially for the deep sea of the Southern Ocean covering 27.9 million km ^{2}. Recent investigations, especially those of the ANDEEP expeditions, revealed an extraordinarily high species richness at abyssal depths. More than 1,400 species of invertebrates were identified (from only the taxa investigated) and more than 700 of these were assumed to be new to science (Brandt et al. 2007a). For example, within protists, the formaminiferan-like komokiaceans were not known from the Southern Ocean deep sea. Now 50 species are reported from that area of which 35 are undescribed (Godday et al. 2007). Within the macrofauna, the isopods were the most diverse taxon with 674 species, of which 87% are putative new species. If we compare these numbers with the more than 4,400 known marine isopod species from the world oceans, the recent Southern Ocean deep-sea expeditions will add approximately 15% to our knowledge on the worldwide zoogeography of that taxon. For the megafauna, the occurrence of new Hexactinellida (glass sponges) and carnivorous demosponges and the first report of Southern Ocean calcareous sponges (Calcarea) were among the most surprising results (Janussen & Reiswig 2009; Rapp et al. in press).

Despite the incomplete faunal knowledge, several studies show linkages between the Antarctic fauna and that of the adjacent deep sea. These studies benefited from a new biologically orientated view on Antarctic seawater temperature. Satellite images show that the well-known Southern Ocean hydrodynamic isolation separating warm surface water in the north from cold water in the south along the Antarctic Convergence is superimposed by horizontal gyres. These allow floating material, for example larvae, other pelagic organisms, pieces of algae, or material serving as substratum for benthic species to penetrate this boundary in both directions (Clarke et al. 2005; Barnes et al. 2006).

Thus, it is mainly the temperature difference that allows only very few species to survive at both sides of the Antarctic Convergence, rather than the front acting as a hydrodynamic barrier. The comparison between the surface and near seabed temperature shows more obviously than ever before how less isolated are the Antarctic bottom-dwelling fauna – including those on the Antarctic shelf from those in the adjacent deep sea. This has relevance not only for future scenarios under climate change but also major implications for the dispersal of animals at evolutionary and ecological timescales.

Hypotheses have always existed about such large-scale dispersal processes. The colonisation of the deep sea by Antarctic organisms seemed to be most likely and most common, after the post-Gondwana breakup and establishment of the ACC. Using genetic techniques, phylogenetic trees can be better linked to plate tectonics, especially the opening of deep-water basins between Antarctica and adjacent continents and the resulting global water mass circulation. Recently, evidence has been provided for an evolutionary dispersal of deep-sea octopods that evolved from common Antarctic ancestors around 30 million years ago into the northerly adjacent deep sea, called tropic submergence (Strugnell et al. 2008). Similar development can be reconstructed for isopods (Asellota, Antarcturidae, Acanthaspidiidae, Serolidae, Munnidae, and Paramunnidae; Raupach et al. 2004, 2009; Brandt et al. 2007b), the amphipod *Liljeborgia*, of which the Antarctic representatives still have eyes whereas their deep-sea relatives are blind (d'Udekem d'Acoz & Vader 2009), and the mollusk *Limopsis* (K. Linse, unpublished observations).

In the opposite direction, multiple evolutionary invasions from the deep sea to the Antarctic shelf, called polar emergence, are very likely for some other isopods, for example Munnopsidae, Desmosomatidae, and Macrostylidae because of their lack of eyes (Raupach et al. 2004, 2009). Similar interpretations are made for representatives of the deep-sea octopod *Benthoctopus* (Strugnell et al. in press). Such examples of long-term evolutionary dispersal have also been described for other taxa such as hexactinellid sponges, pennatularians, stalked crinoids, and elasipod holothurians but have never been studied in detail. Using techniques to decipher the molecular clock, the echinoid *Sterechinus* and the ophiuroid *Astrotoma agassizii* (Hunter & Halanych 2008; Díaz et al. in press) were found to be examples of a split between shallow Antarctic and subantarctic species, which occurred not more than 5 million years ago when glacial–interglacial cycles started.

This was long after Antarctica disconnected from South America and the Antarctic Convergence formed. Similar results are available for the limpet *Nacella* (González Wevar et al. in press) and the bivalve *Limatula*. Perhaps the most extreme example for cryptic speciation is the sea slug *Doris kerguelensis*, from which approximately 29 lineages are derived (Wilson et al. 2009). This puts the development of the Antarctic Convergence 25 million years ago as a main agent of vicariance in question. Surprisingly, this relatively recent split of species within a broad geographical range happened independently of their dispersal potential, because these taxa clearly differ from each other in their early life history traits.

If, despite these few faunistic teleconnections, Antarctica's fauna differs considerably from that of the adjacent slope and the deep sea, for example in the Weddell Sea (Kaiser et al. in press) and from that north of the Antarctic Convergence as for deep-sea gastropods (Schwabe et al. 2007; Schrödl et al.), the reasons must be searched for in polar-

, slope-, or deep-sea-specific environmental parameters. At the level of evolution one major mechanism to generate such biogeographical heterogeneity on the Antarctic shelf is the climate diversity pump, being a modified vicariance concept (Clarke & Crame 1989). Until a few years ago this concept was used to explain a relatively high richness of species with a predominantly circumpolar distribution. It was assumed that during glacial periods populations were spatially separated by grounded ice shelves and as a consequence a radiation of species occurred. At the end of a glacial period when the ice retreated, these new species supposedly mixed around the continent but were obviously not able to interbreed anymore.

This has resulted in sibling species, for example ten sympatric octopods of the genus *Pareledone* (Allcock 2005; Allcock et al. 2007, in press), analoguous to approximately eight cryptic species of the isopod *Ceratoserolis* (Raupach & Wägele 2006) and six allopatric species of *Glyptonotus* (Held 2003; Held & Wägele 2005; Leese & Held 2008; C. Held, unpublished observations). Mostly allopatric cryptic species also occur among the dendrochirote and aspidochirote holothurians, for example among *Laetmogone wyvillethomsoni* and *Psolus charcoti* (Oapos;Loughlin et al. in press) and the amphipod *Orchomene* sensu lato (Havermans et al. submitted). Significant genetic differences have also been found among the pantopod *Nymphon* in the East Antarctic Peninsula and Weddell Sea (Arango et al. in press) and the comatulid crinoid *Promachocrinus* west of the Peninsula and in the Weddell Sea (Wilson et al. 2007) as well as off East Antarctica (L. Hemery & M. Eléaume, unpublished observations). The narcomedusa *Solmundella bitentaculata* was previously thought to be a single ubiquitous species but molecular studies suggest that it contains at least two cryptic species (D. Lindsay et al., unpublished observations).

Resulting from this, a milestone in evolutionary biodiversity research of the past years might be the paradigm shift from an assumed circumpolar macrobenthos to an obviously long-term patchy occurrence of closely related sibling or cryptic species in many taxa.

If, however, the large-scale pattern of the shelf-inhabiting Antarctic macrobenthos is analysed, using the current best available dataset, only one single bioregion is found (Griffiths et al. 2009). The exception is gastropods following the pattern of a split into the Scotian subregion mainly comprising the Antarctic Peninsula and the High Antarctic Province, as proposed by Hedgpeth (1969), which was already questioned a few years later (Hedgpeth 1977). The difference between the interpretations is that the one bioregion result is based on fully reproducible presence/absence datasets with an incomplete systematic coverage. Hedgpeth's conclusion of two provinces included impressions of abundances and consequently of dominance patterns referring mainly to higher taxa and life forms. Additional bias can be caused by the fact that traditional results from the Peninsula were mainly from shallow waters whereas the rest of the Antarctic shelf was sampled at greater depth.

The Southern Ocean Continuous Plankton Recorder (CPR) Survey (Hosie et al. 2003) was the major contribution of the CAML to the research on the Antarctic pelagic system and provided a close link to the Convention on the Conservation of Antarctic Marine Living Resources (CCAMLR). Use of the CPR has significantly increased our knowledge of Antarctic plankton communities by extending the time series and increasing the geographic coverage of the Southern Ocean CPR Survey to approximately 70% of the region, with the highest resolution off East Antarctica. In the 2007/2008 CAML-campaign alone, 15 nations were involved using eight ships conducting 88 successful tows and over 23 transects at 10 m water depth. Since 1991, 25,791 samples have been taken with a resolution of 5 nautical miles, covering a total of 128,955 nautical miles. In terms of large-scale patterns, previous analyses of the Southern Ocean CPR data have shown latitudinal zonation of zooplankton across the ACC, the Sub-Antarctic Front (SAF) acting as a geographic barrier with different species found north and south of it (Hunt & Hosie 2003, 2005). The copepod *Oithona similis* is not only an example for the large-scale pattern but also for temporal changes.

South of the SAF and moving toward the continent, distinct assemblages could be identified which were associated with zones within the ACC. Differences between the assemblages were subtle and based primarily on variation in abundances of species relative to each rather than differences in species composition itself. The CAML provided the opportunity to assess circum-Antarctic patterns. Only night data from the period between December and February were used, rare taxa were excluded, adults and juveniles were merged, and unidentified groups removed. The results on the fauna sampled by the CPR showed no clear longitudinal differences between sectors. In other words, the species composition and abundances of zooplankton within any band of the ACC are effectively the same: it is one community. Tows in January 2008 across Drake Passage did show lower abundances and diversity, but no substantial differences from other transects were observed later in February. The Bellingshausen Sea did show very low abundances and fewer plankton species. The large concentrations of krill, especially in the West Atlantic sector, were not sufficiently covered by this survey. Probably because of the method used, a neritic community only became obvious among the semipelagic, cryopelagic (ice preferring), and pelagic fish (Koubbi et al.), which is dominated west of the Antarctic Peninsula by Antarctic rock cod *Notothenia* and at high Antarctic latitudes by *Trematomus*, Channichthyidae (icefish), and the pelagic *Pleuragramma antarcticum* (O'Driscoll et al. in press). Other planktonic studies embedded in CMarZ used nets with smaller mesh sizes and sampled at greater depth than before. As a consequence, not only were the planktonic fauna more diverse than previously thought, but also many new species were discovered, including the ice-associated fauna.

Microorganisms and the Gelatinous Plankton Likely Belonged to the Most Under

Microorganisms and the gelatinous plankton likely belonged to the most under-represented groups of organisms in Antarctic surveys. During the CAML phase, the understanding of both the extent and ecological variability of Antarctic marine bacterioplankton diversity was greatly enhanced. In just one study approximately 400,000 sequence tags spanning a short hypervariable region of the SSU rRNA gene were determined for 16 samples collected from four regions (Kerguelen Islands, Antarctic Peninsula, Ross Sea, and Weddell Sea) (Ghiglione & Murray, unpublished observations). This effort revealed over 25,000 different sequence tags, of which 13,000 represented equivalents to new species (at a distance greater than 0.03 from the nearest known sequence in public databases). Samples at a low-activity cold seep in the Larsen B area, west of the Antarctic Peninsula, revealed 29 seep-related operational taxonomic units of bacteria and 10 of Archaea, of which 20–30% have no closely cultivated relatives (Niemann et al. 2009). The numbers of gelatinous plankton species increased by a factor of 2–3, especially among hydromedusae, siphonophores, and scyphomedusae, particularly within the neritic assemblage (Lindsay et al., unpublished observations).

Apex predators were also included in the CAML studies. An extensive census in the Atlantic sector of the Southern Ocean, mainly west and east of the Antarctic Peninsula (Scheidat et al. 2007a), showed that whale diversity was higher than expected. Four rare toothed whales from the family of the beaked whales (Ziphiidae) were registered: Arnoux's beaked whale (*Berardius arnuxii*), Gray's beaked whale (*Mesoplodon grayi*), strap-toothed whale (*M. layardii*), and southern bottlenose whale (*Hyperoodon planifrons*), the last with occurrences only in waters deeper than 500 m. Some of the sightings were southernmost records (Scheidat et al. 2007b).

Ecologically Driven Community Heterogeneity between Extremes

One milestone to which CAML researchers contributed is a paradigm shift from a supposed Antarctic circumpolar benthos being rich in species, life forms, and biomass (Figs. 11.5B, C and D) to the general understanding that there is a full range of benthic assemblages from extremely diverse to extremely meager.

Within such a heterogeneous patchwork, poor assemblages were already known decades ago; however, during the CAML phase these were more intensively studied, for example on seamounts (Bowden et al. submitted) and in areas formerly covered by the ice shelf (Gutt et al. in press). This extreme variability can also be attributed to the pelagic system, where on the one hand krill swarms are extremely rich in biomass, but on the other hand extremely low biomass and production are known from the winter season, with a deepest-ever recorded Secchi depth of 80 m, measured on October 13, 1986 in the Weddell Sea (Gieskes et al. 1987). At the seafloor, extremely low abundances can be found in different habitats; at shallow depths with permanent disturbance, in fresh iceberg scours (Gutt & Piepenburg 2003), and under the ice shelf (Gutt 2007). The question of how extremely low abundances can be explained is especially challenging. Unfavourable environmental conditions can lead to the total absence of specific life forms or ecological guilds, such as filter feeders. If food supply is poor then perhaps no more than a few individuals the size of a tennis ball in an area of a tennis court could exist. However, abundances in the formerly ice shelf covered Larsen B area east of the Antarctic Peninsula remained at obviously even lower levels, observed in situ during a Polarstern expedition in 2007, five years after the ice shelf disintegrated (Gutt et al. in press). Because reduced long-term dispersal capacity, at least of species with a circumpolar distribution, can hardly explain this alone, a hypothesis was developed that a poor temporal predictability of food supply during the early life phase could explain extremely rare abundance of adults (Gutt 2007).

Very low biodiversity is also known from different seamounts. At the Admiralty Seamount (East Antarctic), high local densities of stalked crinoids (Hyocrinidae), brachiopods and suspension-feeding ophiuroids (*Ophiocamax*) may reflect ecological conditions such as low predation pressure and low food supply or evolutionary factors (Bowden et al. submitted). The sediment here was dominated by crinoid ossicles, indicating a long persistence of these populations.

In contrast, the benthos of the Scott Seamount less than 400 km away at the same latitude was characterised by a higher abundance of predators, including lithodid crabs, regular sea urchins, and sea stars. A very similar pattern had previously been found on the Spiess Seamount, with large specimens of sea urchins (*Dermechinus horridus*) as well as lithodid crabs (*Paralomis elongata*) being the most conspicuous species and, like the Admiralty Seamount, the seafloor was almost completely covered by spine debris (J. Gutt, unpublished observations).

These differences of dominant species might not only represent temporal parallel ecological processes leading to different results: they could also represent differently advanced stages of long-term developments because stalked crinoids resemble ancient Palaeozoic assemblages and predators indicate a more modern benthos. Generally, in situ images of crinoids could even be used to sample wide-range information on near-bottom current, which is important in explaining benthic community structures (Eléaume et al. in press).

Also, early recolonisation stages of iceberg scours or areas after ice-shelf disintegration can, but do not necessarily always, consist of almost monospecific assemblages such as bryozoans, cnidarians, or ascidians. A dominance of one single species due to an assumed competitive displacement seems to be rare on the Antarctic shelf, but was observed, for example for the sponge *Cinachyra barbata* in the Weddell Sea. Favourable environmental conditions can cause a clear dominance on shallow hard and soft substrata, for example of the scallop *Adamussium colbecki*, the limpet *Nacella concinna* (Barnes & Clarke 1995a; Chiantore et al. 2001), or the infaunal clam *Laternula elliptica* and the deep-sea holothurian *Elpidia glacialis* (Gruzov 1977; Gutt & Piepenburg 1991).

The richness of species of the Southern Ocean deep sea has already been discussed and does not support the hypothesis of a gradient of decreasing richness toward the south (Brandt et al. 2007a). Even the opposite was found for gastropods (Schrödl et al.), which is in contrast to shallow habitats. Communities on the shelf can reach extremely high values for wet weight biomass, up to 12 kg m^{-2} (Gerdes et al. 2003), with relatively high biodiversity compared with the Arctic shallow water. And they are not always defined by the well-known sponge concentrations: recently an assemblage shaped by the elsewhere rare hydrocoral *Errina* has been discovered (CCAMLR 2008).

A high geographical turnover of macrobenthic assemblages within larger regions can generally be explained mainly by sea-ice conditions and proxies for food supply such as current and pigments in the sediment (Gutt 2007; V. Cummings, unpublished observations). Such a regional co-existence of different communities is found almost everywhere, at the Antarctic Peninsula (Lockhart & Jones 2008), at smaller places such as the well-investigated Admiralty Bay including macroalgae (Siciski et al. submitted), in the Weddell and the Ross Seas (Gutt 2007; V. Cummings, unpublished observations), or off East Antarctica (Gutt et al. 2007). For selected deep-sea polychaetes, the challenging question of how allied species can co-exist without out-competing each other was answered by their different food preferences, analysed by biochemical analyses (Würzberg et al.).

Use of the CPR to study plankton patterns has shown that the large-scale zonation around Antarctica (subantarctic, sea ice, neritic) is consistent with latitudinal oceanographic zones as defined by Orsi et al (1995) or Sokolov and Rintoul (2002), Takahashi et al. (2002), Umeda et al. (2002), Hunt & Hosie (2003, 2005, 2006a, 2006b), and Takahashi et al. (2010). These patterns are superimposed by the sea-ice margin and related melting processes, which directly affect the success of some species and consequently the entire community (Raymond & Hosie 2009). The Bellingshausen Sea, for example, exhibited low diversity and abundances. Temporal changes during the past decade have been observed with a decrease in the dominance of krill in the sea-ice zone of eastern Antarctic and an increase in dominance by smaller zooplankton more typical of the permanent open ocean zone, notably the cyclopoid copepod *Oithona similis*, small calanoid copepods *Calanus simillimus* and *Ctenocalanus citer*, foraminiferans, and larvaceans.

Zooplankton abundances in general increased by about 50 times but probably with a lower effect on the total biomass because generally the shift was to small species. Besides the above-described coarse and well-known circumpolar pattern, no latitudinal zonation became obvious within a broad band of the ACC ranging from approximately 52° to 64° S covering a temperature range between a sea surface temperature of 2 and 6 °C. However, within the single surveys differences between predominantly north–south-orientated transects became visible. It is too early to speculate whether the temporal and spatial turnover in community structure is a result of global change, or a shift between two natural events.

The mesopelagic fish fauna, mainly comprising myctophids, were only recently recognised as a key component in the open ocean system because they prey on mesozooplankton, especially copepods, and in turn are the major prey of top predators.

They also contribute to fast vertical energy flux of organic material due to their vertical migrations (Koubbi et al. submitted). Major success has been made in understanding their ecological demands and physical environment, for example in terms of chlorophyll a, sea surface temperature, salinity, and nutrients. Based on that, predictions can be made about suitable habitats for their biodiversity, even for areas from which no such biological data exist.

Small-Scale Heterogeneity, a Contribution to Large-Scale Biodiversity

Epibiotic relationships have become more obvious since the first seabed photographs were taken in the late 1950s. However, for a long period, symbiotic associations, which include parasitic relationships, were judged to be rare in Antarctica (AAVV 1977, p. 389). Later, sponges, bryozoans, and cnidarians (Figs. 11.5B, C and D) were recognised as the main substratum for a variety of echinoderms. In total 347 of such interspecific relationships were found using imaging techniques (Gutt & Schickan 1998). More recent and detailed studies revealed that cidaroid sea-urchin spines alone provide the microhabitat for some 156 species, for example bryozoans, sponges, bivalves, and holothurians, of which some even live obligatorily on the spines. So far, 23 especially close associations (encompassing commensalism, associational defence, and parasitism) have been reported for the Antarctic.

Hosts are generally echinoderms, whereas the symbionts are mainly mollusks and polychaetes (S. Schiaparelli, unpublished observations). The more such symbioses are searched for, the more that are found, for example between a polynoid polychaete and the holothurian *Bathyplotes bongraini* (Schiaparelli et al. in press). Many Antarctic symbioses represent relict interactions, already present before the isolation and cooling of the continent. They might play an important role in explaining an ecological coexistence of species. Such specific relationships are also considered to characterise a mature ecosystem. In the area where the Larsen Ice Shelf recently disintegrated, the composition of epibiotic species did not differ from that living on boulders (Hétérier et al. 2008; Linse et al. 2008; Hardy et al. in press), which indicates a transitional stage of ecological development. A possible hypothesis could be that, in general, symbionts not only linearly contribute to biodiversity as all other species do, but also by providing potential living substrata they might instead accelerate the increase in Antarctic macrobenthic biodiversity by attracting other species.

Applied Aspects and Biodiversity Change

Antarctic waters might be the best protected marine areas on Earth owing to the "Protocol on Environmental Protection to the Antarctic Treaty" (the "Madrid Protocol"). Science managers and politicians have not given high priority to specific marine nature conservation actions for a long time. Recently, interest in such approaches has increased. In 2008 the CCAMLR adopted a proposal by Australia, based on CAML's CEAMARC expedition Collaborative East Antarctic Census of Marine Life (Hosie et al. 2007), to declare two areas of the Southern Ocean mentioned above as Vulnerable Marine Ecosystems (VMEs) because of their complex and vast coralline assemblages. The purpose of the classification is to protect the sites from longline fishing impact, a major concern after bottom-trawl fishing was banned in the most profitable area west of the Antarctic Peninsula (CCAMLR 2008). In addition to several VMEs, one of the world's largest Marine Protected Areas (MPAs) has recently been designated in an area south of South Georgia.

CCAMLR initiated a bioregionalisation project (Grant et al. 2006), which predicts potential habitats for key ecological species and assemblages in order to identify biological hot spots (Koubbi et al.). The United Nations Environmental Program developed criteria to define Ecologically and Biologically Significant Areas as determined by the Convention on Biological Diversity in 2008, which is independent of any sustainable use of the ecosystem. The Scientific Committee on Antarctic Research recently compiled a comprehensive Report on the "Antarctic Climate Change and the Environment" (Turner et al. 2009), which addresses the necessity for baseline information and long-term observations to monitor the mainly climate-induced affects on the ecosystem. All these initiatives were significantly supported, some even initiated, by leading CAML scientists. The results also provide a valuable basis to keep the Red List of Threatened Species updated.

Bioprospection in the Antarctic is still in its infancy. Providing that international law and conventions are respected, CAML can contribute to a further development of this opportunity, and consequently of Antarctic ecosystem services to the benefit of human well-being. Marine biodiversity is protected from large-scale offshore fertilization for CO_2 mitigation by the Convention on Biological Diversity and the Madrid Protocol to the Antarctic Treaty. Fish stocks around the Antarctic Peninsula have been protected from bottom trawling since the 1990–91 season. The development of fish stocks was observed around Elephant Island and the lower South Shetland Islands in December 2006 – January 2007 surveys (K.-H. Kock, unpublished observations). One of the most abundant species before exploitation, the mackerel icefish (*Champsocephalus gunnari*), has not yet recovered. The status of the second target species of the fishery, the marbled notothenia (*Notothenia rossii*), is unclear as no specific surveys for the species have been conducted. Bycatch species, icefish, and nototheniids, appear to have recovered since the area was closed to commercial fishing. Unexplainable so far is the recruitment failure in the past seven or eight years of the yellow notothenia (*Gobionotothen gibberifrons*). The stock currently consists to a very large extent only of adult fish.

Hot and Cold Spots of Biodiversity

During the CAML period, knowledge of Antarctic biogeography steadily increased, in some cases exponentially, in others blank spots were filled. We are now able to identify more local biodiversity hot (and cold) spots. However, regionally comparable criteria are still difficult to apply. In this context it is necessary to establish a systematic geographic coverage, as spatially homogenous as possible for as many as possible relevant regions, and not so much to reach detailed results at one single location. Then, the total number of

CAML has demonstrated that life in the coldest marine ecosystem on Earth is rich, unique, and worthy of high-priority study. We have taken a significant step forward toward the long-term aim of a complete documentation of its biodiversity and a comprehensive understanding of its sculpting forces. The major findings at the evolutionary scale are of a large systematic coverage of cryptic species with distinct non-circumpolar occurrence. Extreme ecological heterogeneity exists at various spatial, biological, and temporal scales. The key to convincing decision-makers to finance a progressive continuation of this work is to assess the contribution of Antarctica's biodiversity to human well-being and ecosystem services. This can include sustainable exploitation of genetic information and the recognition of the role of the ecosystem as a natural CO_2 sink. Antarctica's life is part of the global biodiversity in which causes of and biological response to anthropogenic impact are spatially separated. It is recognised as "the canary in the coal mine", able to provide early warning of dire environmental effects of global warming (IPCC 2007). In this respect, future marine biodiversity surveys have an import.

A Census of Zooplankton of the Global Ocean

The animals that drift with ocean currents throughout their lives (that is, the holozooplankton) include approximately 7,000 described species in 15 phyla. The holozooplankton assemblage is the focus of the Census of Marine Zooplankton, which has produced comprehensive new information on species diversity, distribution, abundance, biomass, and genetic diversity. Our realm among Census of Marine Life projects is the open ocean; we have sampled biodiversity hot spots throughout the world's oceans: little-known seas of Southeast Asia, deep-sea zones below 5,000 m, and polar seas.

We have used traditional plankton nets and newer sensing systems deployed from ships and submersibles. Our analysis has included traditional microscopic and morphological examination, as well as molecular genetic analysis of zooplankton populations and species. CMarZ has contributed to Census legacies in data and information for the Ocean Biogeographic Information System and proven technologies of DNA barcoding. Our photograph galleries of living plankton have captured public interest, and our training workshops have enhanced taxonomic expertise in many countries. The knowledge gained will provide a new baseline for detection of impacts of climate change, and will contribute to our fundamental understanding of biogeochemical transports, fluxes and sinks, productivity of living marine resources, and marine ecosystem health.

Historical Perspective

Despite more than a century of sampling the oceans, comprehensive understanding of zooplankton biodiversity has eluded oceanographers because of the fragility, rarity, small size, and/or systematic complexity of many taxa. For many zooplankton groups, there are long-standing and unresolved questions of species identification, systematic relationships, genetic diversity and structure, and biogeography.

There has never been a taxonomically comprehensive, global-scale summary of the current status of our knowledge of biodiversity of marine zooplankton. Although studies of the taxonomy, distribution, and abundance of zooplankton date back as far as the middle of the nineteenth century, worldwide distribution patterns have not been mapped for all described species. The cosmopolitan or circumglobal distributions characteristic of holozooplankton species of many groups have created special difficulties for accurate biodiversity assessment. The snapshots from different parts of the world ocean have rarely been merged together, in part because the complicated and time-consuming task of compiling the information from numerous individual publications is undervalued.

For most zooplankton groups, significant numbers of species remain to be discovered. This is especially true for fragile (for example gelatinous) forms that are difficult to sample properly and for forms living in unique and isolated habitats, such as the water surrounding hydrothermal vents and seeps (Ramirez-Llodra et al. 2007; Chapter). All regions of the deep sea are certain to continue to yield many new species in multiple taxonomic groups. The practical difficulties of exploring these regions are gradually being overcome, and they are likely to continue to yield new species discoveries for many years.

Our perception of zooplankton biodiversity has almost certainly been affected by their small size, resulting in a marked under-description of species and morphological types. Until recently, some pelagic taxa (for example foraminifers, copepods, euphausiids, and chaetognaths) have been thought to be well known taxonomically, but the advent of molecular genetics has altered this perspective. Morphologically cryptic, but genetically distinctive, species of zooplankton are being found with increasing frequency and will probably prove to be the norm across a broad range of taxa. Many putative cosmopolitan species may comprise morphologically similar, genetically distinct sibling species, with discrete biogeographical distributions. This issue is especially relevant for widely distributed species and/or for species with disjoint distributional ranges, including those occupying coastal environments (Conway et al. 2003). It is likely that many morphologically defined zooplankton species will be found to consist of complexes of genetically distinct populations, but how many cryptic species are present is currently unknown, even for well-known zooplankton groups.

Marine zooplankton are important indicators of environmental change associated with global warming and acidification of the oceans. A global-scale baseline assessment of marine zooplankton biodiversity, including long-term monitoring and retrospective analysis, is critically needed to provide a contemporary benchmark against which future changes can be measured. Knowledge of previous and existing patterns of zooplankton distribution and diversity is useful for management of marine ecosystems and assessment of their status and health (Link et al. 2002). Marine zooplankton are also significant mediators of fluxes of carbon, nitrogen, and other critical elements in ocean biogeochemical cycles (Buitenhuis et al. 2006). Species composition of zooplankton assemblages may have strong impacts on rates of recycling and vertical export; long-term changes in fluxes into the deep sea (Smith et al. 2001) may be related to zooplankton species composition in overlying waters (Roemmich & McGowan 1995; Lavaniegos & Ohman 2003).

Compared with the dimensions of the known – in terms of numbers of species and regions of the world's oceans – the unknown is thought to be many times larger. Introducing his monograph on the biogeography of the Pacific Ocean, McGowan (1971) posed several questions that help frame the unknown territory of zooplankton biodiversity. "What species are present? What are the main patterns of species distribution and abundance? What maintains the shape of these patterns? How and why did the patterns develop?" Nearly 40 years later, the answers to these questions remain poorly known for many ocean regions and most zooplankton groups.

Zooplankton Sampling

Zooplankton samples for CMarZ have been collected by nets, buckets, water bottles, sediment traps, light traps, remotely operated vehicles (ROVs), submersibles, and divers. Sampling strategies have trade-offs for each type of sampling gear: some may obtain numerous specimens, but under-sample fragile taxa, whereas others may be suited for collecting fragile organisms for taxonomic analysis, but may be unable to sample at spatial resolutions and scales appropriate for accurate characterisation of patterns of distribution and abundance.

During CMarZ dedicated cruises in the Atlantic Ocean, zooplankton and micronekton were quantitatively sampled throughout the water column using MOCNESS (Multiple Opening/Closing Net and Environmental Sensing System; Wiebe et al. 1985; Wiebe & Benfield 2003). In addition to collecting depth-stratified plankton samples, the MOCNESS transmits environmental data (depth, temperature, salinity, horizontal speed, and volume filtered) to the ship throughout the tow; the data are recorded for subsequent analysis. A uniquely equipped 10 metre MOCNESS allowed CMarZ to sample to 5,000 m in the Atlantic Ocean and rapidly filter large volumes (tens of thousands of cubic metres) to capture rare deep-sea zooplankton (Wiebe et al. 2010). The collections included first-ever observation of living specimens of rare deep-sea species, and offered remarkable opportunities for photographing living specimens and barcoding novel species.

CMarZ has used modern in situ survey technologies, including crewed submersibles, ROVs, towed camera arrays, and visual/video plankton recorders (VPR; Davis et al. 1992) to observe and collect zooplankton, especially fragile gelatinous forms, in many areas of the ocean. These sampling approaches have led to new species discoveries (Haddock et al. 2005; Lindsay & Miyake 2007), and rapid advances in our understanding of deep sea biology and ecology (Pagès et al. 2006; Ates et al. 2007; Fujioka & Lindsay 2007; Kitamura et al. 2008a, 2008b; Lindsay et al. 2008). In 2006, Dhugal Lindsay (Japan Agency for Marine-Earth Science and Technology) led a pilot study to census gelatinous and hard-bodied zooplankton in Sagami Bay (Japan) using diverse sampling technologies, including an autonomous video plankton recorder (AVPR) with a high-definition video camera for colour imagery. The study yielded images and samples of zooplankton and marine snow that are being analysed to model and predict effects of climate change on carbon cycling and sequestration.

Blue-water SCUBA diving for observing and collecting fragile zooplankton was developed during the past 30 years (Hamner 1975), and has been used to advantage by CMarZ. A group of divers work from an inflatable boat launched from a research vessel; they are connected to a central line by a 10 metre tether line and overseen by a safety-diver. This technique has proven ideal to locate, observe, photograph, and collect live and undamaged specimens of free-swimming gelatinous animals.

A variety of remote plankton-sensing platforms (that is, those deployed from ships that return data – but not necessarily samples) has been developed for the study of zooplankton diversity, distribution, and abundance. CMarZ has used several among the many instruments developed for this purpose, including the video plankton recorder (VPR; Davis et al. 1992); underwater video profiler (UVP; Gorsky et al. 1992, 2000); optical plankton counter (OPC; Herman 1988); and continuous plankton recorder (CPR; Hardy 1926; Glover 1962). In general, these systems provide higher spatial resolution than nets and more accurate depiction of the animal in its environment (Mori & Lindsay 2008). For those species that can be remotely identified, these instruments are valuable tools in describing the geographical and temporal changes in zooplankton populations in relation to behaviour and the environment.

To census the world's oceans, CMarZ has used ships of opportunity to sample zooplankton in open-ocean waters and areas not regularly frequented by large research vessels. Ships of opportunity have deployed ROVs and crewed submersibles, which usually require large ocean-going vessels for their deployment, in studies in Monterey Bay, California (Matsumoto et al. 2003; Raskoff & Matsumoto 2004) and off the coast of Japan (Lindsay et al. 2004, 2008; Kitamura et al. 2005; Lindsay & Hunt 2005; Lindsay & Miyake 2009). In particular, the Plankton Investigatory Collaborative Autonomous Survey System Operon (PICASSO) ROV system was designed for deployment from ships of opportunity to study gelatinous plankton as deep as 1,000 m (Yoshida et al. 2007a, 2007b; Yoshida & Lindsay 2007).

Sample Preservation

Zooplankton samples for CMarZ have been processed as bulk unsorted samples, especially during cruises of opportunity, and as individual expertly identified specimens, usually during dedicated CMarZ surveys. No single sampling-handling approach can preserve the appearance and morphological, molecular, and biochemical properties of zooplankton specimens. CMarZ developed and has used a sample-splitting protocol that entails immediate bulk processing of a portion of the sample (partly in formalin for morphological analysis and partly in alcohol for molecular analysis), with another portion retained alive for photography, observation, and identification of living specimens, some of which may not be suitable for eventual preservation. Splitting is not recommended for samples with few individuals or rare species, but may in other cases optimise sample use among scientists. Samples for molecular analysis were preserved in 95% non-denatured ethanol or buffer solution (for example RNAlater) and then stored at low temperatures (–20 °C) to slow degradation. Identified specimens were flash-frozen in individual vials in liquid nitrogen. Overall, best results were obtained when DNA extractions were done very soon after collection.

Sample Analysis

An essential element of CMarZ has been traditional morphological examination of samples by taxonomic experts, who are essential to validate species identifications for uncertain and possible new species, examine and confirm putative new or cryptic species, and describe new species. Such skills are the domain of a very few specialists worldwide and are a diminishing resource. The lack of manpower – both expert and technical – has been a bottleneck for CMarZ in our progress toward our goal of a global, taxonomically comprehensive biodiversity census.

Consequently, CMarZ has championed integrated morphological and molecular genetic approaches to analysis of zooplankton species' diversity. A revolution in the analysis of global patterns of species diversity has been driven by the widespread use of DNA barcodes (that is, short DNA sequence used for species recognition and discrimination; Hebert et al. 2003). The usual barcode gene region for metazoan animals is a 708 base-pair region of mitochondrial cytochrome oxidase I, mtCOI (Schindel & Miller 2005). CMarZ barcoding efforts have included analysis of both targeted taxonomic groups and particular ocean regions or domains. Five CMarZ barcoding centres (at the University of Connecticut, USA; Ocean Research Institute, Japan; Institute of Oceanology, China; Alfred Wegener Institute, Germany; and National Institute of Oceanography, India) have worked together toward a shared goal of determining DNA barcodes for the approximately 7,000 described species of zooplankton. CMarZ has also uniquely demonstrated the use of off-the-shelf automated DNA sequencers in shipboard molecular laboratories, allowing a continuous at-sea analytical "assembly line" from collection, identification, and DNA barcoding.

Environmental DNA surveys (that is, determination of sequences for 16S or 16S-like rRNA coding regions from mixed environmental samples) have transformed our understanding of microbial diversity in the oceans (Pace 1997; Sogin et al. 2006). CMarZ has applied this revolutionary approach to the analysis of zooplankton species diversity based upon COI barcodes, using an approach dubbed environmental barcoding (that is, DNA sequencing of the COI barcode region from unsorted bulk samples). This approach has the marked advantage of not requiring morphologically based species identification. For zooplankton, environmental barcoding entails comparison of the resultant DNA sequences with "gold standard" DNA barcode data to identify species and characterise species diversity (Machida et al. 2009)

Data and Information Management

CMarZ uses a centralised distributed data and information management system, an outgrowth of the US GLOBEC Data and Information Management System (Groman & Wiebe 1998; Groman et al. 2008), which integrates among three primary data centres: Woods Hole Oceanographic Institution (Woods Hole, USA), Ocean Research Institute (Tokyo, Japan), and Alfred Wegener Institute (Bremerhaven, Germany). The ready and open exchange of information helps ensure that CMarZ project participants can coordinate and avoid duplication of effort, and thus speed progress toward the goal of a comprehensive and complete DNA barcode database for zooplankton

Toward a Global View of Pelagic Biodiversity

Compared with the approximately 1 million described terrestrial insects and more than 1 million benthic marine organisms, the diversity of marine zooplankton, with about 7,000 species, is by no means rich. A unique attribute of this assemblage is the relative magnitude of local diversity to global diversity (Angel et al. 1997). As an example, the Copepoda – the most species-rich group of marine zooplankton – are very common and species are frequently very abundant. One net sample from oceanic waters may contain hundreds of copepod species or about 10% of the global total of approximately 2,200 species. This ratio is nearly unique among animal groups and habitats.

Low global diversity has been attributed to the homogenous and unstructured pelagic environment compared with terrestrial, intertidal, or benthic habitats. High local diversity has been attributed to the coexistence of many species, through vertical or other modes of niche partitioning, but the exact mechanism for their co-existence is still poorly understood (Lindsay & Hunt 2005; Kuriyama & Nishida 2006). Recently, the contribution of biological associations toward the enhancement of species diversity has been attracting much attention (Pagès et al. 2007; Lindsay & Takeuchi 2008; Ohtsuka et al. 2009).

Since 2004, CMarZ has completed more than 90 cruises, and samples for CMarZ have been collected at more than 12,000 stations; an additional 6,500 archived samples have been available for analysis. CMarZ has sampled from every ocean basin. For selected groups of zooplankton, CMarZ has made excellent progress toward a new global view of biodiversity. Although zooplankton are not as prevalent as microbes, for which an "everything is everywhere" debate continues, species with circumglobal distributions are found in every phylum of the zooplankton assemblage from Protista to Chordata. Such broadly distributed species have been a focus of particular attention for CMarZ. The global biogeography of planktonic Foraminifera has been mapped by Colomban de Vargas (CNRS, France), based upon integrated morphological and molecular systematic analysis (de Vargas et al. 2002; Morarda et al. 2009). Demetrio Boltovskoy (University of Buenos Aires, Argentina) has produced an atlas of Radiolaria (Polycystina) distributions based upon 6,719 samples that reveals relations between radiolarian distributions and worldwide water mass and circulation patterns (Boltovskoy et al. 2003, 2005). CMarZ contributed to production of a monograph on the known genera of Hydrozoa in the world ocean (Bouillon et al. 2006). Analysis of global patterns of copepod diversity and abundance is being performed by Sigrid Schnack-Schiel (Alfred Wegener Institute, Germany), who is comparing regional patterns in tropical, temperate, and polar seas; in all regions, more than 50% of all species occur in low abundances (not more than 10 individuals per 100 cubic metres); in Antarctic waters, more than 80% of species are rare.

CMarZ has also contributed to new understanding of ocean-basin scale patterns of species diversity through monographic treatments of selected zooplankton groups. Notable among these are analyses of planktonic Ostracoda of the Atlantic Ocean (Angel et al. 2007; Angel 2008; Angel & Blachowiak-Samolyk 2009; Angel 2010). Also, Vijayalakshmi Nair (National Institute of Oceanography, India) has advanced understanding of species diversity of the Chaetognatha, a taxonomically challenging group, in the Indian Ocean (Nair et al. 2008) and, working with Annelies Pierrot-Bults (University of Amsterdam, The Netherlands), in the Atlantic Ocean. Gelatinous zooplankton diversity patterns have been found to differ between the Pacific Ocean and Japan Sea sides of Japan (Lindsay & Hunt 2005), including unique investigations of ctenophores and other fragile gelatinous zooplankton using submersibles below 2,000 m (Lindsay 2006; Lindsay & Miyake 2007). An in-depth study on the gelatinous fauna of the Gulf of Maine was published by Pagès et al. (2006). Also, checklists and field guides have been produced to aid in species identification of gelatinous plankton for Japanese waters (Lindsay 2006; Kitamura et al. 2008a, 2008b; Lindsay & Miyake 2009); for waters off California (Mills et al. 2007; Mills & Haddock 2007); and for the Mediterranean (Bouillon et al. 2004).

Biodiversity Hot Spots

Sampling within regions and/or for taxa that have historically been ignored or understudied has been a key objective of CMarZ. Our efforts have been focused on biodiversity hot spots (that is, geographic or taxonomic domains for which there is greatest scope for improved knowledge of species richness), which may be specific areas of the ocean, taxonomic groups, or ecological guilds. Marine ecologists and oceanographers must identify and priorities such regions, similar to terrestrial ecologists, who have identified 18 biodiversity hot spots based primarily on degree of endemism and impacts of human activities (Wilson 1999).

Among the numerous acknowledged biodiversity hot spots for marine zooplankton, CMarZ has focused on diversity in the deep sea, polar seas, and coastal regions and marginal seas of Southeast Asia. Our taxonomic targets have included gelatinous groups and other taxonomically challenging and under-appreciated groups throughout the zooplankton assemblage. The CMarZ focus on geographic and taxonomic areas with high potential for species discovery has resulted in discoveries of 89 new species, of which 52 have been formally described.

Southeast Asian Coastal Waters and Marginal Seas

Comprehensive research has been conducted in the embayed waters, coastal areas, and marginal seas of Southeast Asia. This is a major biodiversity hot spot in the world and has a very complicated geography and geological history. New species discoveries here have been dominated by copepods (including *Pseudodiaptomus*, *Tortanus* (*Atortus*), and species of the families Pontellidae and Pseudocyclopidae) and mysids collected using sledge nets from coastal near-bottom habitats and by night-time or SCUBA sampling in coral reefs, indicating that the high diversity of these habitats has been overlooked by conventional daytime net sampling (see, for example, Nishida & Cho 2005; Murano & Fukuoka 2008). The Sulu Sea, a semi-enclosed marginal sea in the tropical Western Pacific Ocean, has been a particular focus for CMarZ studies (Nishikawa et al. 2007) and has yielded several discoveries of new species and genera, including copepods (Ohtsuka et al. 2005).

Species discoveries by CMarZ within the Copepoda have added another 8% to the total number of copepod species in Southeast Asia (another 2% to the global total), and new species discoveries of Mysidacea in Southeast Asia have added 15% to the global total for that group. Understanding the significance of these numbers must also take into account the ecological importance of the species and their role in the ecosystem. Regardless, CMarZ has made exceptional progress in improving our knowledge of zooplankton biodiversity in Southeast Asia by building effective teams of expert taxonomists who collaborate with CMarZ scientist

The Deep Sea

By volume, 88% of the ocean environment is deeper than 1 km and 76% is between a depth of 3 and 6 km. The deep sea is thus the largest habitat on earth – and also the one least known. Previous studies have yielded several general characteristics of pattern of zooplankton diversity, distribution, and abundance in the deep sea. A primary finding is that numbers of species and their abundances tend to decrease with depth (Longhurst 1995). The decrease in number of species is not linear; there is a peak in mid-water layers and a decrease with greater depth. However, better sampling in the deep sea and discovery of new species at depth may alter this trend. Latitude affects this general trend, with higher numbers of species at all depths in lower latitudes than at higher latitudes (Angel 2003). Other general trends are that deeper-dwelling species are less likely to be endemic (that is, native and restricted to a particular region) and more likely to be geographically widespread. Usual feeding mode varies through the depth strata, with filter-feeding herbivorous species occurring in the upper water layers, and detritivores and carnivores most abundant below the light-filled surface waters.

Exploration and discovery in the deep sea have been slowed by the inherent difficulties of sampling at great depths. Deeper than 1,500 m in most ocean areas, the very low abundances of most species requires that huge volumes of water be filtered, with sampling over many hours using huge sampling systems deployed from large ships, to collect significant numbers of individuals. Despite these challenges, many new deep-sea species have been discovered in the past decade, strongly indicating that deep-sea biodiversity has so far been markedly underestimated.

CMarZ' unique approach to sampling deep-sea zooplankton using a 10 metre MOCNESS with fine mesh nets has yielded many discoveries and first-time observations of living specimens. During two CMarZ cruises using this gear to explore the deep tropical/subtropical Atlantic Ocean regions (that is, the Sargasso Sea on the R/V RH *Brown* in 2006, and the eastern Atlantic on the FS *Polarstern* in 2007), zooplankton were collected from the entire water column with a focus on describing species composition and richness and discovering new species in the poorly known meso and bathypelagic zones. The Sargasso Sea cruise yielded a treasure-trove of specimens, including Ctenophora (22 species), Cnidaria (110 species), Ostracoda (58 of 140 known Atlantic species), Copepoda (134 species), euthecosome pteropod Mollusca (20 of approximately 33 species), heteropod Mollusca (17 of

29 species), Cephalopoda (13 species), and Appendicularia (13 of approximately 70 species). In addition, 3,965 fish specimens were collected, including 127 species of 84 genera from 42 families. Below 1,000 m depth, the MOCNESS-10 collected several little-known species, including the siphonophores *Nectadamas richardi* (Pugh 1992) and *Lensia quadriculata* (Pagès et al. 2006).

During the eastern Atlantic cruise, more than 1,000,000 cubic metres of seawater was filtered and approximately 60,000 specimens were identified. In some cases, collections represented a significant fraction of the species known from the South Atlantic; 104 copepod species were identified of an estimated total of 500 species known (Bradford-Grieve et al. 1999). A sample from the bathypelagic captured a putative new copepod species, the third to be described from the family Hyperbionychidae (Bradford-Grieve 2010). From the two CMarZ deep-sea Atlantic cruises, at least 15 novel ostracod species were discovered and are in process of description (Martin Angel, unpublished data).

In recent years, CMarZ' use of *in situ* sampling and observation from submersibles and ROVs has dramatically improved our understanding of deep-sea biodiversity, biology, and ecology. Laurence P. Madin (Woods Hole Oceanographic Institution, USA) led a CMarZ exploration to the Celebes Sea, a tropical sea and biodiversity hot spot in the Indonesia/New Guinea/Philippine triangle between the Pacific and Indian Oceans. Sampling was performed by blue-water diving and net systems; deep-sea observations to 3,000 m used a Global Explorer ROV with high-definition television and benthic-baited video "Ropecams".

The team discovered that the overall biomass of the water column was high, with exceptional abundance of the nitrogen fixing, blue-green bacteria *Trichodesmium*. Sperm whales and spinner dolphins were observed at the surface, squid were seen from the ROV, and myctophid fishes were collected in the trawl. Ten of 23 known worldwide species of Salpidae, a group of gelatinous zooplankton, were collected by blue-water divers. Two species thought to be new to science were observed: a black, benthopelagic lobate ctenophore and a large pelagic polychaete worm with ten long cephalic tentacles.

Further CMarZ deep-sea exploration using ROVs and submersibles uncovered a cascade of biological associations of the deep-sea hydromedusan, *Pandea rubra*, which is dependent upon a pteropod mollusk for the polyp stage of its life cycle (Lindsay et al. 2008). Ocean acidification is thought to be detrimental to calcareous shell-bearing Mollusca, and the newly discovered linkage between these species may represent a threat to the medusa. *Pandea rubra* was found to host many other species during its deep-sea medusa stage, including Pycnogonida, hyperiid Amphipoda, and larval stages of other hydromedusae (Lindsay et al. 2008). Invaluable archived video from the 11,000 m ROV *Kaiko* revealed what appeared to be a new order of Ctenophora in the Ryukyu Trench (Japan); a comb jelly was observed floating above and attached by "strings" to the sea floor at a depth of 7,217 m (Lindsay & Miyake 2007).

Polar Seas

As a general rule across pelagic groups, species diversity is lower at high latitudes than at low latitudes. Although the explanation for this remains unclear, low temperature and dramatic seasonal shifts in light levels and sea ice cover – and thus primary production – surely represent significant challenges to survival. Although the most characteristic feature of polar seas is sea ice, early studies of polar zooplankton were largely restricted to ice-free areas and summer months. This has severely limited our understanding of polar ecosystems, because the sea ice environment is a unique environment harbouring a diverse fauna (Bluhm et al. 2010) and plays a vital role in ecosystem dynamics of both polar oceans.

In the Antarctic, where sea ice is predominantly seasonal, the Southern Ocean krill (*Euphausia superba*) is the keystone species and inhabits the seasonal pack-ice zone of Antarctic Coastal Current (Atkinson et al. 2004; Siegel 2005). Copepoda are dominant in many Antarctic regions in terms of both biomass and abundance, with few large species (for example *Calanus propinquus*, *Calanoides acutus*) making up more than 40% of total copepod biomass, and frequently neglected smaller species (for example *Oithona*, *Oncaea*, *Microcalanus*, *Ctenocalanus*, and others) accounting for more than 80% of total copepod abundance (Kosobokova & Hirche 2000; Hopcroft & Robison 2005; Schnack-Schiel et al. 2008). Park & Ferrari (2008) reported a total of 205 calanoid copepod species from the Southern Ocean: 184 species (of which 50 are endemic) were restricted to deep waters, 13 species (8 endemic) were epipelagic, and 8 species (all endemic) were neritic.

The Arctic Ocean is unique owing to its permanent and seasonal ice cover, and restricted exchange of deep-water biota with the Pacific and Atlantic Oceans. Extreme environmental conditions and limited exchange with the adjacent ocean regions have resulted in a zooplankton assemblage comprising species endemic to the Arctic Ocean and uniquely adapted to cold temperatures (Smith & Schnack-Schiel 1990; Kosobokova & Hirche 2000; Deibel & Daly 2007). Approximately 300 species of holozooplankton have been recorded for the Arctic (Sirenko 2001).

The greatest diversity occurs within the Copepoda (approximately 150 species), which dominate the zooplankton community in both abundance and biomass (Kosobokova & Hopcroft 2009). Four large calanoid species (*Calanus glacialis, C. hyperboreus, C. finmarchicus,* and *Metridia longa*) are by far the most dominant species, contributing 60–70% of total zooplankton biomass. Cnidaria are represented by approximately 50 species, mostly hydromedusae; mysids contribute approximately 30 species, most of which are epibenthic. Other groups are each represented by fewer than a dozen described species.

Gelatinous Zooplankton

Special attention has been paid to the biodiversity of gelatinous plankton as a hot spot for species discovery. Discoveries of novel Cnidaria and Ctenophora species have resulted, some requiring the establishment of new higher taxonomic groups (Lindsay & Miyake 2007). This work has also allowed comparisons among regional faunas in light of geological history and environmental conditions, and revealed novel relationships among gelatinous plankton and other organisms.

DNA Barcoding, CMarZ has championed integrated morphological and molecular genetic approaches to analysis of zooplankton species' diversity. Importantly, CMarZ has placed a high priority on "gold-standard" barcoding (that is, determination of a 500+ base-pair DNA sequence for mtCOI for an identified vouchered specimen, with specified metadata for protocols and collections) for described species of zooplankton. The growing CMarZ barcode database – now approaching 2,000 species or about 30% of the approximately 7,000 described species – will serve as a "Rosetta Stone" for species identification of marine zooplankton, linking species names, morphology, and DNA sequence variation. DNA barcodes will thus facilitate rapid characterisation of patterns of species diversity and distribution in the pelagic realm.

Taxon-specific barcoding efforts by CMarZ researchers have included analysis of every phylum and taxonomic group within the zooplankton assemblage, including Protista (Morarda et al. 2009); Cnidaria (Ortman 2008; Ortman et al. 2010), calanoid Copepoda

(Machida et al. 2006), Euphausiacea (Bucklin et al. 2007), Ostracoda (Angel et al. 2008), pteropod Mollusca (Jennings et al. 2010a), Chaetognatha (Jennings et al. 2010b), among other groups. These studies have demonstrated the usefulness of DNA barcodes for identification of known species, discovery of new species, and recognition of cryptic species within widespread or poorly known taxa. Alternatively, CMarZ barcoding campaigns have had a regional focus, with barcoding of all identified specimens collected from a particular ocean region or during a survey cruise. CMarZ has regionally focused barcoding efforts either completed or ongoing in the Arctic Ocean (Bucklin et al. 2010a), Sargasso Sea (Bucklin et al. 2010b), Eastern Atlantic and the Bay of Biscay, and South China Sea.

DNA barcodes have been used to characterise large-scale patterns of population genetic diversity and structure (that is, the amount and distribution of genetic variance within and among natural populations of organisms). Global-scale sampling and molecular analysis of zooplankton has revealed geographically distinct and genetically differentiated populations of Copepoda (Blanco-Bercial et al. 2009; Machida & Nishida 2010); Euphausiacea (Bucklin et al. 2007); and Chaetognatha (Peijnenburg et al. 2004; Miyamoto et al. 2010). Although geographic populations of zooplankton are not reproductively isolated, they are important units of evolution; studies of population genetic diversity and structure, using the barcoding gene region of choice, have been a critical aspect of the CMarZ global biodiversity assessment.

CMarZ has uniquely demonstrated the feasibility and value of performing DNA barcoding at sea during oceanographic research cruises. During two CMarZ biodiversity surveys to the Sargasso Sea and the Eastern Atlantic, DNA extraction, polymerase chain reaction (PCR), and DNA sequencing were performed in shipboard barcoding laboratories. Hundreds of species were barcoded, based on specimens identified by the taxonomic experts also participating in the cruise. During the Sargasso Sea cruise, 329 DNA barcodes were determined for 191 holozooplankton species, including hydrozoans, crustaceans, chaetognaths, and mollusks; barcodes were determined for an additional 35 fish species (Bucklin et al. 2010b).

CMarZ has pioneered the use of environmental barcoding for zooplankton communities: Machida et al. (2009) sequenced the COI barcode from an unsorted bulk sample collected in the western equatorial Pacific Ocean, detected 189 species of zooplankton based on COI sequences, and demonstrated the usefulness of this powerful approach to estimating species diversity of metazoan animals.

As the DNA barcode database has grown to include described species collected from diverse ocean regions, CMarZ has explored new approaches for computationally efficient analysis and useful presentation of DNA sequence data and results. A novel approach is vector analysis (Sirovich et al. 2009), scalable analysis for very large datasets that produces heuristic displays of barcode similarity called heat maps or because of their resemblance to modern art – "Klee diagrams".

Patterns of Historical Change

Time-series observations and monitoring programs that span many years are needed to document long-term changes in zooplankton diversity, distribution, abundance, and biomass in ocean ecosystems (Perry et al. 2004). CMarZ has sought to embed our field activities in the context of such valuable programs in order to provide benchmark biodiversity information for the analysis of temporal changes associated with global climate change.

Northeast Atlantic Ocean, UK

Since 1946, the Continuous Plankton Recorder (CPR) Survey Program (Sir Alister Hardy Foundation for Ocean Science, UK) has sought to characterise and understand changes in the species composition of North Atlantic zooplankton. Approximately 2.5 million non-zero records have indicated that zooplankton biomass has declined below the long-term average. In the North Sea, present biomass levels are one-half those in 1960 and warm-water species (for example the copepod *Calanus helgolandicus*) are displacing cold-water species (for example *C. finmarchicus*). This change affects the survival of fish larvae that depend on *C. finmarchicus* and has far-reaching consequences for the ecosystem (Beaugrand et al. 2002; Edwards et al. 2007). Another finding from analysis of CPR data is evidence of seasonal shifts in spawning and other life-history processes related to climate warming, creating mismatch between fish larvae and their food (Edwards & Richardson 2004; Edwards et al. 2008) and resulting in low fish recruitment. CPR surveys are beginning to document trans-Arctic migrations, with unknown consequences for pelagic communities (Edwards et al. 2008). CPR data have revealed what are called "regime shifts" (that is, markedly increased diversity and decreased productivity of zooplankton) in the North Sea (Edwards et al. 2007), as well as the Northwest Atlantic, Northeast Pacific, and Northwest Pacific Oceans (Kane & Green 1990; Pershing et al. 2005; Kane 2007; Mackas et al. 2007; Tian et al. 2008)

Northwest Atlantic Ocean, USA

A 40 year survey by the US National Marine Fisheries Service (NMFS) has used the CPR (Jossi & Goulet 1993) to correlate effects on zooplankton populations with decadal scale forcing by the North Atlantic Oscillation (Greene & Pershing 2000; Conversi et al. 2001; Piontkovski et al. 2006; Turner et al. 2006). Analysis of data and samples collected during 1977–2004 by NMFS' Marine Resources Monitoring, Assessment and Prediction (MARMAP) program (Jossi & Kane 2000; Kane 2007), demonstrated that total zooplankton counts over Georges Bank were at or above long-term average levels between 1989 and 2004 (Kane & Green 1990; Kane 2007). This regime shift was also observed in CPR records from the Gulf of Maine, which showed an increase in smaller-sized species (Pershing et al. 2005; Greene & Pershing 2007). The US GLOBEC Northwest Atlantic Program/Georges Bank Study examined zooplankton dynamics during 1994–1999, and provided one of the most comprehensive species datasets available for an historical fishing ground (Wiebe et al. 2002). Since 2004, NMFS has provided samples to CMarZ from quarterly ecosystem monitoring surveys over the Northwest Atlantic continental shelf; these samples are being used to barcode 200 of the most abundant zooplankton species, with a goal of allowing rapid DNA-based biodiversity assessments for fisheries management in this area.

Benguela Current, South Africa

Zooplankton diversity and biomass have been recorded for the Benguela Current, west of South Africa, since 1951. In contrast to most Eastern Boundary Currents, numerical abundances have increased 100 fold over recent decades. Species composition has shifted, with smaller-bodied Copepoda and Cladocera now dominating in the region, perhaps because of the combined effects of changes in climate, circulation patterns, and predator dynamics (Verheye & Richardson 1998; Verheye 2000).

California Current, USA

The California Cooperative Oceanic Fisheries Investigations (CalCoFI) is a 60 year time-series of quarterly surveys off southern and central California. No long-term trend was detectable in total zooplankton carbon biomass, although zooplankton displacement volume has declined in both regions, likely due to declines in salp abundances (which are mostly water and contribute little to carbon biomass; Lavaniegos & Ohman 2007). Zooplankton biomass varied among areas along the coast (Fernandez-Alamo & Färber-Lorda 2006).

The time-series showed a marked shift from a "warm" regime with low zooplankton biomass and high sardine populations to a "cool" regime with high zooplankton biomass and high anchovy populations

Western Pacific, Japan

The Odate collection at the Tohoku National Fisheries Research Institute (Japan) contains 20,000 formalin-preserved zooplankton samples collected throughout the western North Pacific from 1950 to 1990. Analysis of this extensive collection revealed decadal oscillations in copepod diversity and abundance. Compared with all other decades, the late 1980s showed a climatic regime shift: the copepod *Metridia pacifica* was dominant, whereas the abundance of *Neocalanus plumchrus* was low (Sugisaki 2006; Tian et al. 2008).

Indian Ocean, India

The International Indian Ocean Expedition (IIOE), carried out during 1962–1965, was the most intensive sampling program to characterise zooplankton species diversity and abundance in the region. Digitization and analysis of the IIOE data by CMarZ scientists have characterised seasonal variability associated with monsoon conditions and revealed long-term trends in species abundances and biogeographical distributions (Nair et al. 2008; Nair & Gireesh 2010). The data and results from IIOE represent an invaluable resource for detecting seasonal, interannual, and long-term shifts in the zooplankton assemblage (Baars 1999).

Marginal Seas

In the Caspian Sea, the invasive comb jelly *Mnemiopsis leidyi* has severely impacted the entire ecosystem since its introduction in the late 1990s (Kideys et al. 2005, 2008). In the Black Sea, 130 years of plankton records (1870–2000) documented a long-term increase in the diversity of tintinnid Ciliata until 1960, followed by a decline into the 1990s. Ctenophore blooms in the 1990s may have impacted the anchovy fishery by causing replacements of copepod species (Gavrilova & Dolan 2007) and generating a trophic cascade with disastrous consequences for the ecosystem (Kideys et al. 2005).

Arctic Ocean

Changes in zooplankton species diversity, distribution, and abundance can be expected to occur in the Arctic Ocean, as climate change alters water temperature and thus timing and magnitude of productivity cycles (Grebmeier et al. 2006; Bluhm & Gradinger 2008). The introduction of exotic species through increased commercial traffic and the establishment of expatriate species from adjoining regions with warming are likely for the Arctic Ocean in the near future (Acia 2004). Comparisons of zooplankton abundances in recent years with the early 1950s suggest that some taxa were more abundant in the Chukchi and Beaufort shelf and slope regions and the Canada Basin in 2002 than about 50 years ago (Grebmeier et al. 2006

Taxonomic Training

CMarZ has enhanced research capacity in marine biodiversity and zooplankton ecology through the training of graduate students by CMarZ scientists and through international exchanges of students, staff, and researchers among CMarZ laboratories. CMarZ has placed a high priority on training new zooplankton taxonomists, with a total of 252 participants for 27 Taxonomic Training Workshops during 2004–2009. Many students have joined the CMarZ Network, which now includes more than 150 members, to seek information and access to expertise to facilitate their research. CMarZ has sought to enhance capacity for taxonomic analysis of zooplankton through both traditional morphological and molecular systematic analysis

Applications of CMarZ Results

CMarZ' efforts toward a global assessment of marine zooplankton biodiversity, focusing on geographic and taxonomic hot spots, will provide a benchmark baseline biodiversity assessment for measurement of future changes resulting from climate change or other anthropogenic or natural variation. Zooplankton diversity can also be used as a measure of the health of marine ecosystems, and knowledge of prior and existing patterns of zooplankton distribution and diversity is needed for the management of coastal marine ecosystems (Link et al. 2002). Zooplankton are pivotal players in the dynamics of marine ecosystems, and new knowledge is needed of their roles in biogeochemical cycles (Buitenhuis et al. 2006).

CMarZ' barcode database, protocols for barcoding diverse marine phyla, and techniques for environmental sequencing of zooplankton will be useful in accelerating analysis of zooplankton diversity and distribution for a variety of applications in ocean research, management, and conservation. DNA barcodes will be used to produce DNA microarray "chips" for automated and/or remote identification and quantification of zooplankton. In the not-too-distant future, ocean-observing stations may include moored instruments with DNA-based detection systems for in situ species identification. These same approaches may allow rapid and accurate species identification for ecosystem monitoring and fisheries management, detection of invasive species in ballast water, and other possibilities central to ocean observing, management, and regulation.

Challenges and Opportunities for the Future

The question of how many species are present still remains, and future studies will continue to challenge our understanding of diversity. Large-scale studies of zooplankton are needed to evaluate patterns of biodiversity at scales appropriate to dispersal ability in ocean currents. Shifts in geographic ranges may underlie apparent temporal changes observed during spatially limited studies. In the case of species introductions, clear definition of potential source populations and likely colonisation pathways require an understanding of global-scale distributions. For some cosmopolitan species, there may be little genuine endemism, whereas others may consist of complexes of genetically distinct entities, representing geographically isolated populations or cryptic species.

There remain large gaps in the sampling coverage done by CMarZ, especially in the deep sea and under-sampled ocean regions. Despite the increase of deep-sea studies, these investigations still represent rare snapshots of this huge habitat and vast areas of the deep sea remain unexplored. Despite new capabilities of ice-breaking research vessels, our understanding of polar ecosystems during the dark season remains fragmentary. Despite our efforts, CMarZ has only performed limited sampling in the Central Pacific Ocean, obtaining some samples from ships of opportunity (for example sailing ships associated with the Sea Education Association, Woods Hole, USA). Comprehensive sampling in the Indian and tropical Pacific Oceans has not yet extended below a few hundred metres. The richly diverse benthopelagic communities have received almost no attention because of sampling difficulties.

The urgency and/or vulnerability of regions or taxa to anthropogenic or natural threats must be weighed in determining future research priorities. These include regions where rates and impacts of climate change are most likely to be amplified, and poorly studied areas threatened by anthropogenic inputs (such as near population centres in emerging nations). The availability of baseline data is a critical issue, because evaluation of biodiversity patterns and hot spots requires improved knowledge of existing data and trends. Sites where time-series collections or long-term monitoring studies have been done are of high priority for continued assessment.

An essential feature for continued progress toward a global zooplankton census will be an international partnership, ideally coordinated through a network of regional centres that can identify opportunities for cooperative field work, arrange sampling from ships of opportunity, and lead efforts to secure funding for dedicated cruises. Interdisciplinary collaborations – among oceanographers, ecologists, taxonomists, geneticists, geochemists, and others – will continue to be needed to answer increasingly complex questions and address increasingly critical issues about the future and health of the global ocean. Knowledge of the species diversity, distribution, and abundance of the zooplankton assemblage will continue to be a critical element for monitoring, understanding, and predicting the complex global system.

The effort, enthusiasm, and expertise of the CMarZ Steering Group members are the basis of the results reported here. We acknowledge the many contributions by the CMarZ Steering Group members who are not listed as authors herein. They are the following: Martin Angel (National Oceanography Centre, UK); Demetrio Boltovskoy (Universidad de Buenos Aires, Argentina); Janet M. Bradford-Grieve (NIWA, New Zealand); Rubén Escribano (Universidad de Concepción, Chile); Erica Goetze (University of Hawaii, USA); Steven Haddock (Monterey Bay Aquarium Research Institute, USA); Steve Hay (Fisheries Research Services Marine Laboratory, UK); Russell R. Hopcroft (University of Alaska – Fairbanks, USA); Ahmet Kideys (Institute of Marine Sciences, Turkey); Laurence P. Madin (Woods Hole Oceanographic Institution, USA); Webjørn Melle (Institute of Marine Research, Norway); Vijayalakshmi R. Nair (National Institute of Oceanography, India); Mark Ohman (Scripps Institution of Oceanography, USA); Francesc Pagés (deceased) (Institute of Marine Sciences, Barcelona, Spain); Annelies C. Pierrot-Bults (University of Amsterdam, The Netherlands); Philip C. Reid (Sir Alister Hardy Foundation for Ocean Science, UK); Song Sun (Institute of Oceanology, Chinese Academy of Sciences, China); Erik V. Thuesen (The Evergreen State College, USA); Colomban de Vargas (Roscoff Marine Station, France); Hans M. Verheye (Marine & Coastal Management, South Africa). We acknowledge the support of the Alfred P. Sloan Foundation. This study is a contribution from the Census of Marine Zooplankton an ocean realm field project of the Census of Marine Life.

7

The Future of Marine Animal Populations

The Census of Marine Life's overarching goal is to assess and explain the diversity, distribution, and abundance of marine organisms throughout the world's oceans. By stimulating exploration and research in all ocean habitats it has accumulated an unprecedented wealth of new information on the patterns and processes of marine biodiversity on a global scale. Three questions are guiding this research effort. What did live in the oceans? What does live in the oceans? What will live in the oceans? The Future of Marine Animal Populations (FMAP) Project ultimately aims to answer that third question through the analysis and synthesis of available data, and the modelling of patterns and trends in marine biodiversity. This entails all levels of biodiversity, from individuals, to populations, communities, and ecosystems (Box 16.1).

FMAP engaged primarily in the statistical modelling of ecological patterns derived from empirical data. The emphasis has been on data synthesis, often by means of meta-analysis, which is the statistical integration of multiple datasets to answer a common question (Cooper & Hedges 1994). FMAP researchers have also engaged in field surveys and experimental work, but have mostly focused on analysing and synthesizing datasets collected by other Census projects and third parties. This approach enabled us to ask broad scientific questions about the status and changes in diversity, abundance, and distribution of marine animals, such as the following:

- What are the global patterns of biodiversity across different taxa?
- Which are the major drivers explaining diversity patterns and changes?

- What is the total number of species in the ocean (known and unknown)?
- How has the abundance of major species groups changed over time?
- What are the ecosystem consequences of fishing and other human impacts?
- How are animal ranges and their distribution in the ocean changing?
- How is the movement of animals determined by behaviour and the environment?

The main limits to knowledge have been missing data on species that have not been counted, mapped, or tagged, and in some cases missing access to existing data on species that have been monitored. From a statistical perspective, the main challenge has been to overcome data limitations such as the limited length of most time series, the problem of temporal or spatial autocorrelation, and separating ecologically relevant patterns from environmental noise and measurement error.

Despite the ultimate focus on future prediction, the synthetic analyses undertaken within the FMAP project inform all three aspects of the Census, past, present, and future. The rationale is that without a solid understanding of past and present trends, it is impossible to make sound future projections. Likewise, our research efforts encompass different levels of organisation, from the movements of individual animals through space and time, to broad macro-ecological patterns of abundance and diversity. Hence, an improved understanding of processes at the level of an individual animal may help inform the interpretation of larger-scale patterns. Our main analytical tools are meta-analytic models, used to combine and understand species abundance and distribution trends, including both historical and recent data. Models that are effective for synthesis also have potential for prediction, and have been used by others to project potential future effects of fishing and climate change, for example Botkin et al. (2007). Moreover, modelling can help define the limits of knowledge: what is known and how firmly, what may be unknown but knowable, and what is likely to remain unknown in the foreseeable future.

FMAP grew out of a workshop held at Dalhousie University in Halifax, Nova Scotia, Canada, in June 2002. Representatives of other Census projects, including the History of Marine Animal Populations (HMAP) project, the field projects, and the Ocean Biogeographic Information System (OBIS), participated and provided guidance in the design of this project. FMAP was originally envisioned and led by Ransom A. Myers, Killam Chair of Ocean Studies at Dalhousie University. His leadership carried the project until his sudden passing in 2007. Two additional FMAP centres were established in 2003 at the University of Iceland with Gunnar Stefansson, and the University of Tokyo with Hiroyuki Matsuda. Since 2007 the project has been co-led by the authors of this chapter.

FMAP's mission has been to describe and synthesize globally changing patterns of species abundance, distribution, and diversity, and to model the effects of fishing, climate change, and other key variables on those patterns. This work has been performed across ocean realms and with an emphasis on understanding past changes and predicting future patterns. The project benefitted throughout from close collaboration with statisticians and mathematical modelers, which enabled the proper processing and analysis of large datasets. FMAP has collaborated with other Census projects to varying degrees, most consistently with HMAP, Tagging of Pacific Predators (TOPP), and OBIS, as well as various deep-sea projects.

This chapter does not intend to provide an exhaustive overview of the research activities within FMAP for individual projects and publications). Instead, we aim to highlight key areas of interest and discuss major advances that have been made. It is structured along three major research topics, aiming to cover the major research themes of the Census (distribution, abundance, and diversity of marine life): (1) marine biodiversity patterns and their drivers, (2) long-term trends in animal abundance and diversity, (3) distribution and movements of individual animals. In the concluding section we aim to provide some insight into what is unknown, and what is currently unknowable, particularly with respect to predicting the future of marine biodiversity

Biodiversity Patterns and Their Drivers

Before the Census, mapping of the ocean with respect to our knowledge of fundamental patterns of abundance and diversity was limited. The first global study was published in 1999, presenting a pattern of planktonic foraminiferan diversity derived from the analysis of a large sediment core database (Rutherford et al. 1999). Another study highlighted global hot spots of endemism and species richness for corals and associated organisms (Roberts et al. 2002). Several authors had investigated latitudinal gradients for particular species groups (Hillebrand 2004). Yet compared with our understanding of life on land, synthetic knowledge on marine biodiversity was sparse.

It became clear from these early studies, however, that some of the patterns were uniquely different from those seen on land, where biodiversity is generally highest in the tropics (Gaston 2000).

Large Marine Predators

FMAP studies have mainly focused on large pelagic predators such as tuna and billfish, whales, and sharks, for which global data were available. These species groups were found to peak in diversity in the subtropics, often between 20–30 degrees latitude north or south. Although a similar distribution pattern was first described for Foraminifera (Rutherford et al. 1999), we were able to show that this is a more general pattern that applies across very different species groups (Worm et al. 2003, 2005). Furthermore, it became clear that this biodiversity pattern is not static, but dynamically changing on both short and long time scales.

Species richness patterns for tuna (Thunnini), billfish (Istiophoridae), and swordfish (Xiphiidae) were derived from a global Japanese longline fishing dataset. Pelagic longlines are the most widespread fishing gear in the open ocean, and are primarily used to target tuna and billfish. The Japanese data represents the world's largest longline fleet and the only globally consistent data source reporting species composition, catch and effort for all tuna, billfish, and swordfish. Statistical rarefaction techniques were used to standardise for differences in fishing effort and to estimate species richness (the expected number of species standardised per 50 randomly sampled individuals) for each 5° × 5° cell in which the fishery operated. Species richness of tuna and billfish displayed a global pattern with large hot spots of diversity in all oceans in the 1960s. These hot spots faded over time, indicating declining species richness, a pattern most clearly seen in the Atlantic and Indian Oceans.

Declining species richness coincided with 5 to 10 fold increases in total fisheries catch of tuna and billfish in all oceans, which may have led to regional depletion of vulnerable species (Worm et al. 2005). In the Pacific, however, initial losses of diversity began to reverse in 1977, coinciding with a large-scale climate regime shift, whereas the Pacific Decadal Oscillation changed from a cool to a warm phase. Climatic drivers were also found to be important on an annual scale. Short-term (year-to-year) variation in species richness showed a remarkable synchrony with the El Niño Southern Oscillation (ENSO) index, with increasing temperatures leading to basin-wide increases in species richness (Worm et al. 2005). This may be explained by warming of sub-optimal temperature habitats. ENSO related decreases in diversity were seen in the tropical Eastern Pacific, a region that suffers from greatly reduced productivity and associated mass mortality of marine life during El Niño events. A subsequent study showed that seasonal variation in sea surface temperature is driving the taxonomic richness patterns for deep-water cetaceans (whales and dolphins) as well (Whitehead et al. 2008).

For tuna and billfish, as well as cetaceans and Foraminifera, mean sea surface temperature (SST) clearly emerged as the strongest single predictor of diversity, showing a positive correlation over most of the observed temperature range (5–25 °C), but a negative trend above that. This decline of diversity at high temperatures was most pronounced in the western Pacific "warm pool", which has the highest equatorial SST (warmer than 30 °C), and weakest in the tropical Atlantic, which has the lowest equatorial SST (lower than 27°C). The relation between tuna and billfish diversity and SST could also be independently reconstructed from an analysis of individual species temperature preferences (Boyce et al. 2008).

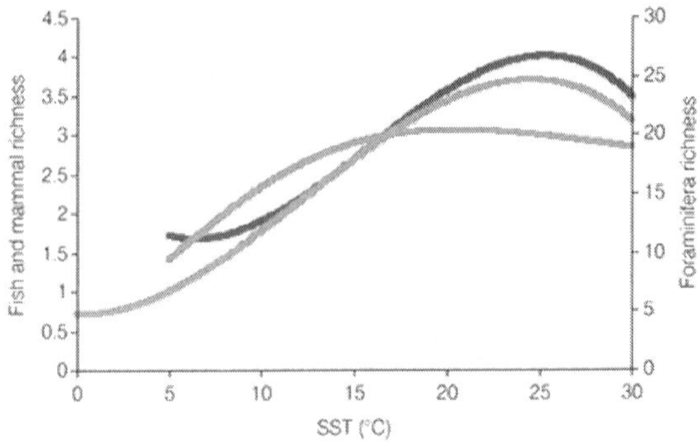

Another factor that explained significant variation in tuna and billfish species richness on a global scale was the steepness of horizontal temperature gradients. Sharp temperature gradients are found around frontal zones and eddies that are typically associated with mesoscale oceanographic variability. Fronts and eddies often attract large numbers of species, likely because they concentrate food supply, enhance local production, and increase habitat heterogeneity (Oschlies & Garçon 1998; Hyrenbach et al. 2000). They may also form important landmarks along transoceanic migration routes (Polovina et al. 2001). Finally, dissolved oxygen concentrations were positively correlated with diversity. This likely relates to species physiology, as low oxygen levels (less than 2 ml l^{-1}) may limit the cardiac function and depth range of many tuna species (Sund et al. 1981). Regions of low oxygen are located west of Central America, Peru, West Africa, and in the Arabian Sea. Despite optimal SST around 25 °C, most of these areas showed conspicuously low diversity.

Knowledge of the relation between SST and diversity for various species groups allows us to predict how diversity may change as SST changes spatially and temporally with climate variability and climate change. The effects of climate variability, such as ENSO and the Pacific Decadal Oscillation, are discussed above. With respect to long-term climate change, Whitehead et al. (2008) combined Intergovernmental Panel on Climate Change (IPCC) scenarios for observed and projected changes in SST between 1980 and 2050 with an empirically derived relation of SST and deep-water cetacean diversity. For the baseline 1980 dataset, diversity was predicted to be highest at latitudes of about 30°, falling towards the equator, and more precipitously towards the poles. With global warming, these bands of maximal diversity were predicted to move pole-wards. The warming tropical oceans were predicted to decline in diversity, while richness was predicted to increase at latitudes of about 50°–70° in both hemispheres (Whitehead et al. 2008). These general conclusions were recently corroborated by an analysis of 1,066 exploited fish and invertebrate species (Cheung et al. 2009).

Other Species Groups

Other groups that were investigated with respect to their diversity patterns were deep-water corals and tropical reef fish. The goal was to gain a better understanding of the effects of human impacts such as fishing and ocean acidification on the distribution, abundance, and diversity of different species groups (reviewed by Tittensor et al. 2009b).

A study on tropical reef fish at fished and unfished sites in three oceans revealed predictable changes in the species–area relation (SAR). The SAR quantifies the relation between species richness and sampling area and is one of the oldest, most recognised patterns in ecology. Fishing consistently depressed the slope of the SAR, with the magnitude of change being proportional to fishing intensity (Tittensor et al. 2007). Changes in species richness, relative abundance, and patch occupancy contributed to this pattern. It was concluded that species-area curves can be sensitive indicators of community-level changes in biodiversity, and may be useful in quantifying the human imprint on reef biodiversity, and potentially elsewhere (Tittensor et al. 2007). This study highlighted how human impacts can affect biodiversity through multiple pathways. Subsequent work focused on cold-water scleractinian corals, an important habitat-forming group of stony corals commonly found on seamounts (Clark et al. 2006).

Despite their widely accepted ecological importance, records of cold-water corals are patchy and simply not available for most of the global ocean. In an FMAP-CenSeam (Global Census of Marine Life on Seamounts) collaboration, the probable distribution of these corals was derived from habitat suitability models, that incorporated all the available data on cold-water coral distribution in relation to environmental variables such as depth, temperature, and carbonate availability (Tittensor et al. 2009a).

Highly suitable habitat for seamount stony corals was predicted to occur in the North Atlantic, and in a circumglobal strip in the Southern hemisphere between 20° and 50° S and at depths shallower than around 1,500 m. Seamount summits in most other regions appeared less likely to provide suitable habitat, except for small near-surface patches. In these models oxygen and carbonate availability played a decisive role in determining large-scale scleractinian coral distributions on seamounts (Tittensor et al. 2009a). These results raise concerns about the possible consequences of ocean acidification (Orr et al. 2005) and the observed shallowing of oxygen minimum zones in the wake of global climate change (Stramma et al. 2008). Both factors would be predicted to limit the distribution of scleractinian corals, and the fauna associated with them.

Total Species Richness

The number of species is the most basic index used to measure biodiversity and one that plays a fundamental role in the quantification of human-related extinctions and impacts. Unfortunately, the total number of species remains poorly known in the oceans. For example, Grassle & Maciolek (1992) famously suggested that the number of (largely unknown) deep-sea benthic species is more than 1 million, but may even exceed 10 million. The only published estimate of the total number of marine species relied on an inventory of European fauna that was scaled up to the global level (Bouchet 2006). A more analytical approach has recently become possible through the Census' Ocean Biogeographical Information System (OBIS) in combination with newly developed modelling approaches (Mora et al. 2008). These modelling methods derive estimates of species richness from "discovery curves" of species sampled over time, and produce confidence limits that allow us to estimate the known and unknown of global species richness. An FMAP pilot project on total marine fish species has estimated that there are approximately 16,000 known species of marine fish, with about another 4,000 awaiting discovery (Mora et al. 2008). These methods are currently being used to estimate the known and unknown of total marine species richness.

Long-Term Trends in Abundance

Underlying the changing patterns of biodiversity or species richness are changes in the abundance and distribution of individual populations. Most previous work has emphasized variability in population abundance in relation to climate, oceanography, or other factors on yearly to decadal or evolutionary time scales (Vermeij 2004; Jackson & Erwin 2006). Changes in marine life over the Anthropocene (the past few hundred years; an epoch dominated by human influences) have only recently received focused attention. This has two reasons: first, the ocean has long been seen as a vast frontier, where human activities would not leave a permanent mark; second, empirical monitoring data are mostly available just for the past 20 to 50 years, which prevented longer-term studies from reaching back beyond the twentieth century.

Synthesizing Long-Term Trends

Over the past decade, the Census at large, and HMAP and FMAP in particular, have partly overcome these limitations. Although HMAP has made enormous progress in unravelling detailed historical, archaeological, and paleontological records of past changes in different animal populations and regions, FMAP has developed ways of combining and analysing these data to reveal long-term changes in ocean ecosystems, and uncover their drivers and consequences.

One of FMAP's goals has been to synthesize the long-term trends in the abundance, distribution, and diversity of marine life. This has been pursued for coastal regions over the past centuries and millennia (Lotze & Milewski 2004; Lotze et al. 2005, 2006) and continental shelf and open ocean regions over the past 50 years (Myers & Worm 2003, 2005; Worm et al. 2005, 2009). These studies have shown that human impacts have resulted in sharply reduced abundance of target and some non-target populations, as well as range contractions and local extinctions that precipitated local and regional losses of species diversity.

To synthesize long-term trends in population abundances of large marine animals, we analysed 256 records from 95 published studies, many of them from HMAP, FMAP, or other Census projects (Lotze & Worm 2009). Trend estimates for marine mammals, birds, reptiles, and fish were derived from archaeological, historical, fisheries, ecological, and genetic studies and revealed an average decline of 89% (range: 11–100%) from historical abundance levels (Lotze & Worm 2009).

Remarkably, the magnitude of depletion was relatively consistent across different species groups despite considerable variability in data quality, analytical methods, and time span of the records. Diadromous fish such as sturgeon and salmon, sea turtles, pinnipeds, otters, and sirenia showed the strongest declines with more than 95%. On the other hand, conservation efforts in the twentieth century enabled several whale, pinniped, and coastal bird species to recover from a historical low point in abundance. These recoveries have reduced the level of depletion across all 256 analysed species to 84% on average.

Another important dimension of change is the spatial expansion of exploitation, which began in rivers and along the coasts centuries ago and only in the mid-twentieth century moved towards open oceans and the deep sea. Thus, some of the highest population declines can be found in rivers and coastal habitats, with lesser declines found on continental shelves and the open ocean. Deep-sea habitats differ from this trend, which may be explained by their extreme vulnerability to exploitation (Roberts 2002).

Along with this spatial expansion there has been a temporal acceleration in exploitation due to technological advances. Population declines unfolded over hundreds or thousands of years in many rivers and coastal regions, one to two hundred years on the continental shelves, approximately 50 years in the open ocean, and approximately 10–20 years in the deep sea (Lotze & Worm 2009). As a result, the average magnitude of change is almost independent of when exploitation started. Interestingly though, recoveries are mostly found in species that have been exploited at least 100 years ago and protected in the early to mid-twentieth century, whereas more recently exploited species do not yet show recovery.

Changes in population abundance and distribution have resulted in changes in species diversity. As was discussed previously, there have been remarkable changes in tuna and billfish species richness in the open ocean over the past 50 years. In the coastal ocean, changes in diversity have occurred in two ways: (1) diversity declines have occurred in large marine animals such as mammals, birds, reptiles, and fish due to global, regional, or ecological extinctions; (2) diversity increases have occurred through the invasion of mostly smaller species including invertebrates, plants, unicellular plankton and bacteria, and viruses (Lotze et al. 2005, 2006). This shift in diversity from large- to small-bodied animals has resulted in a different species composition, with consequences for ecosystem structure, functioning, and services.

Drivers of Long-Term Change

The underlying drivers of observed long-term changes may include natural and anthropogenic drivers, as well as the cumulative effects of multiple factors. To unravel the relative importance of different drivers on population and ecosystem changes, we have used a variety of methods, including meta-analysis of large datasets, experimental manipulations, and ecosystem models.

For example, an analysis of drivers of long-term population changes in 12 estuaries and coastal seas revealed that exploitation (primary factor) and habitat loss (secondary) were by far the most important causes for the depletion and extinction of marine species over historical time scales (Lotze et al. 2006). Pollution, physical disturbance, disease, eutrophication, and introduced predators also contributed to some species declines, although to a lesser extent. However, the reverse was also true: conservation efforts in the twentieth century, especially reduced exploitation, the protection of habitat, and in some cases pollution control, enabled several species to recover from low abundance. In many cases it was not a single factor, but a combination of exploitation, habitat loss, and other factors that caused a population decline or – in reverse – enabled recovery (Lotze et al. 2006). Whereas species invasions and climate change were less dominant drivers of marine biodiversity change in the past, they may increase in importance in the future (Harvell et al. 2002; Harley et al. 2006; Worm & Lotze 2009).

The cumulative and interactive effects of different drivers have also been explored with multi-factorial laboratory experiments. For example, a three-factorial experiment that used rotifers as a model system showed additive effects between exploitation and habitat fragmentation on population declines and synergistic effects if environmental warming was also involved (Mora et al. 2007). Although each of these three factors individually caused populations to decline by similar amounts, all factors combined resulted in up to 50 times faster declines of experimental populations. These results highlight the importance of multiple human drivers for past and future population and biodiversity changes in the ocean. Many human drivers have their ultimate roots in the social conditions and economic activities of society.

Although some individuals, communities, or societies may exercise overexploitation, habitat destruction, and waste dumping, others promote successful stewardship and governance through harvest regulations, pollution controls, and the protection and restoration of species and habitats (Lotze & Glaser 2009; Worm et al. 2009). In one study, Mora (2008) analysed socioeconomic and environmental databases together, to separate the proximate and ultimate drivers of coral reef degradation. Most of the local and regional variability in fishes, corals, and macroalgae was explained by human-related factors such as agricultural land use, coastal development, overfishing, and climate change. Significant ecological interactions among the different species groups further highlighted the need for a comprehensive management of human influences on coral reef ecosystems (Mora 2008).

Ecosystem Consequences

What are the consequences of long-term population changes on the structure and functioning of marine ecosystems today and in the future? And how will changes in ecosystem structure affect the services marine ecosystems provide for human well-being? These questions have been difficult to tackle. First, ecosystem structure, functioning, and services are not easy to quantify and the relevant data are not readily available. This problem was overcome by compiling and analysing long-term datasets on ecological, environmental, and socioeconomic changes in marine ecosystems. Second, a multitude of species, environmental drivers, and human impacts may interact in ways that are impossible to unravel from simple trend analyses. We therefore used ecosystem modelling approaches to determine overall changes in food web structure, energy flows, and stability. These modelling techniques may also serve to project future scenarios under changing environmental or human conditions, which we are currently exploring.

Long-term changes in population abundance as well as the loss (extinction) and gain (invasion) of species has changed the structure of many coastal ecosystems. This can have important effects on ecosystem functioning. Many species fulfill important ecological functions, including the provision of spawning, nursery, and foraging habitat by wetlands, underwater vegetation, and reef-building organisms. Most of these habitats also play an important role in filtering particles, nutrients, and pollutants, thereby maintaining good water quality. If the functioning of coastal ecosystems is compromised, so are the ecosystem services provided for human well-being (Worm et al. 2006; Lotze & Glaser 2009). For example, overexploitation, habitat loss, and pollution have depleted many fisheries that previously provided food and employment. The loss of filter functions together with increasing municipal and industrial discharges have posed health risks to people through harmful algal blooms, contaminants, and disease. Finally, the expansion of oxygen-depleted zones, invasive species, and flooding compromises recreation and shoreline safety (Worm et al. 2006). As human impacts spread offshore and expand to other ocean regions (Halpern et al. 2008), the observed changes in coastal oceans may forecast potential future changes in other habitats.

To understand better the ecosystem effects of marine biodiversity change, we used two different modelling approaches, stochastic network models and mass-balance food web models. First, basic food and interaction webs are assembled from data on species occurrence, abundance, feeding links, and other ecological information (Coll et al. 2008). Next, the two different models enabled us to analyse changes in up to 22 food web properties reflecting (among others) species composition, food-chain length, energy transfer between trophic levels, and the linkage density and complexity of the webs. For example, food webs in the Adriatic and Catalan Seas in the Mediterranean were found to be very similar in terms of their structure and functioning, but were more ecologically degraded compared with food webs from the Caribbean, Benguela, and US continental shelf (Coll et al. 2008). Food web properties estimated by both models yielded very similar results, thereby enhancing confidence in our results compared with any single modelling approach.

The network models also allowed us to analyse the robustness of food webs to simulated species loss (Coll et al. 2008). It had previously been shown that the removal of highly connected species in the food web results in a much higher rate of secondary extinctions than randomly deleted or less strongly connected species. In our analyses, removing the commercial species caused intermediate rates of secondary extinctions indicating that commercial species are often well connected in the food web. Finally, we could show that a larger degree of ecological degradation in the Mediterranean food webs resulted in a diminished robustnes to species loss and a higher rate of secondary extinctions. This suggests that the degradation of marine ecosystems may accelerate the rate of biodiversity loss in the future.

Animal Movements

Movement patterns and behaviour of individual animals collectively contribute to broader-scale population distribution, species' ranges, and patterns of biodiversity. The development of a statistical toolbox for studying the movements of electronically tagged marine predators and collaboration with animal trackers provides a powerful complement to the global-scale studies of biodiversity and abundance conducted by FMAP.

Our aim has been to elucidate the underlying mechanisms that determine animal distributions at the individual scale, and how these contribute to marine predator distribution and biodiversity patterns at broader scales. Much of our previous knowledge on marine animal movement patterns has been inferred from observations of species' departures and arrivals at geographically disparate locations (Carr 1986) or from traditional mark–recapture methods (Hilborn 1990). These studies were necessarily coarse in scale and yielded little insight into the interactions between foraging or migrating animals and their environment. Furthermore, these approaches only revealed movements to locations where observers were present, biasing estimates of movement rates and, more generally, our understanding of marine animal movement patterns.

All this has changed with the introduction of electronic tagging and telemetry technologies that revolutionised our view of animal movement patterns and distribution in the ocean (Block et al. 2001; Birdlife International 2004; James et al. 2005). Satellite and light-based geolocation tracking technologies now allow us to follow marine animals for protracted periods as they make their living in the ocean.

The veritable explosion of tracking studies using increasingly refined technologies has revealed, for example, extraordinary 64,000 km round-trip migrations by sooty shearwaters (Shaffer et al. 2006), physiological mechanisms underlying niche expansion in salmon sharks (Weng et al. 2005), and previously unknown return migrations in white sharks (Bonfil et al. 2005).

Despite these and many other success stories, our ability to track and document animal movements has far out-stripped our ability to conduct sophisticated analyses of rapidly amassing tracking data. This gap proved to be a fertile area of investigation for FMAP.

Statistical Tools and Key Results

The project has played the leading role in developing a state-of-the-art statistical toolbox for electronic tracking data (Jonsen et al. 2003, 2005). This has been a multidisciplinary venture, involving biologists, ecological modellers, and statisticians. Our approach has focused on state-space models (SSMs), which are statistical time-series tools that, in the present context, allow one to estimate true animal positions from error-prone tracking data. State-space models have two components, a process model that describes how animals move from one position to the next and an observation model that relates the true, unobserved positions to the empirical tracking data.

In general, there can be two goals to fitting a SSM to tracking data. The first goal is to estimate the true positions of a tagged animal by accounting for the observation error inherent in the tracking data. This is called state filtering and yields a set of position estimates (and associated uncertainties) that occur over regular time intervals . This kind of filtering is fundamentally different from traditional travel rate filters (McConnell et al. 1992) as all observed locations are modeled by a known probability distribution with implausible observations being down-weighted rather than discarded. The result is that all information contained in the data is used to estimate the true positions.

Breed et al. (2006) used the state filtering approach to gain insight into the sexual segregation of seasonal foraging in adult grey seals breeding on Sable Island, Nova Scotia. Grey seals are an important generalist predator in the Scotian Shelf ecosystem, with a population that has experienced exponential growth over the past 35 years. From October to December and February to March, males used areas along the continental shelf break, whereas females used midshelf regions. Breed et al. (2006) suggested that this broad-scale segregation may help individuals maximise fitness by reducing intersexual competition during primary foraging periods.

The second goal of fitting a SSM is to estimate biological parameters or behavioural states specified in the process model, thereby allowing inference of unobservable processes that drive the movement and distribution patterns. The ability to construct biologically meaningful models and to estimate their parameters directly from complex, error-prone data is the most compelling facet of the SSM toolbox. For analyses of individual movement datasets, this approach is best achieved when individual datasets are combined meta-analytically (Jonsen et al. 2003). Meta-analysis facilitates synthesis of multiple datasets, enabling inference both within and among datasets, and improves parameter estimation from limited datasets.

State-space models allow researchers to think about questions that have no conventional solution. Jonsen et al. (2006) highlighted this by showing that endangered leatherback turtles migrating throughout the North Atlantic slow down at night, perhaps to feed on macrozooplankton that migrate toward the ocean surface and/or because their navigation abilities are less precise than during the day. One problem with this is that calculating travel speeds becomes difficult when travel distance is small during a single day or night period (no more than 30 km) compared with the uncertainty in the observed positions (up to 250 km). Conventional analyses of these data are not able to reveal the patterns in day versus night travel rates that the SSM analysis can. Furthermore, Jonsen et al. (2006) showed that a Bayesian meta-analytic SSM, a model that estimates day versus night travel rates simultaneously from all datasets, yielded superior estimates for individual turtles and provided the basis for prediction at a population level.

A particularly compelling application of the SSM allows one to infer the (hidden) behavioural state of animals based upon the shifts in movement patterns observed in the tracking data. To derive, for example, the probability of an animal foraging versus migrating, a switching model can be added to the standard SSM (Jonsen et al. 2005). Key to this approach are the underlying assumptions that animal movements can be modelled by a small set of correlated random walks, each corresponding to a unique behavioural state, and that animals typically engage in area-restricted type movements (such as slow travel rates with a high frequency of turning) when searching for and consuming prey (Jonsen et al. 2007). The switching model is used to estimate the probabilities that an animal is in a particular behavioural state, conditional upon the previous behavioural state. This approach offers a powerful tool to infer animal behaviours from remote tracking data and is being used by FMAP collaborators to analyse migration and foraging patterns in Pacific leatherback turtles (Bailey et al. 2008; Shillinger et al. 2008). Similar approaches have been adopted for analysing foraging behaviours of southern bluefin tuna (Patterson et al. 2009).

FMAP researchers have used the switching SSM approach to identify potential foraging areas in space and time, a first step in quantifying critical foraging habitat and in developing mechanistic predictions of the potential influence of climate variability and changing distribution of foraging predators. Breed et al. (2009) show that adult male and female grey seals forage in different areas of the Scotian Shelf but both sexes tend to focus on relatively small, intensely used foraging sites. This switching SSM anaysis builds on previous efforts that suggested grey seals foraged over broad areas of the shelf (Breed et al. 2006) by resolving spatial patterns in movement behaviours hidden within the tracking data (Breed et al. 2009). Similar patterns of concentrated foraging activity, albeit at a much broader spatial scale, are emerging in an ongoing analysis of leatherback turtle satellite data. It presents only a small, but fairly representative, subset of the results so far.

Switching SSM analysis of 36 individual tracks, spanning a period from 1999 to 2006, suggests that foraging activity is concentrated along the slope waters of the Scotian Shelf and around Cape Breton in Northeastern Nova Scotia. Both regions are highly productive and likely support large seasonal aggregations of the leatherback's prey, jellyfish

What is striking about both of these analyses is the distinct patchiness of foraging activity exhibited by both species. Grey seals are benthic foragers and are predominantly tied to shallow banks on the continental shelf, whereas leatherback turtles forage on jellyfish both in the coastal and pelagic realms. Regardless, both species show distinct and annually predictable preferences for relatively discrete regions. Understanding how these behavioural patterns may change in the future as a result of changes in prey distribution and abundance, and as a result of climate variability can provide valuable insight into mechanisms underlying future change in species biodiversity and abundance. Current work is focusing on the biophysical characterisation of species' foraging areas and development of switching SSMs that can directly relate environmental gradients to the behavioural patterns hidden within electronic tracking data.

Concluding Remarks

The FMAP Project has been active from 2002 to 2010 as the modelling component of the Census. Our emphasis has been on data synthesis to reveal broad patterns of diversity, abundance, and distribution of marine animals, with a particular emphasis on large predators. We have developed data analysis and modelling techniques that enabled us to estimate the diversity of known and unknown species in the ocean, to derive long-term trends, short-term dynamics, and spatial patterns of diversity change, and to understand better the distribution and behaviour of individual species.

In our communications we have strived to use the knowledge gained to inform society about both the drivers of biodiversity change, as well as current and possible future consequences of biodiversity loss for marine ecosystems and human society. Major results and patterns derived from this project are listed in Box 16.2. In interpreting this body of new knowledge, it is important to realise that these patterns do not play out equally everywhere. A necessary weakness inherent in large-scale data synthesis is that individual differences may be lost in averaging across a large population of animals, different regions, and environmental conditions.

It may be these differences, however, existing between one place and another, or one population and another, that hold the key to understanding the ocean's future (Worm et al. 2009). There is no doubt that, given our large and growing influence on the marine environment, future societal decisions will drive the trajectory of change in marine animal populations and the ecosystems in which they are embedded. Within FMAP we can highlight past and current trends, and assume different scenarios as to how these may extend into the future. What will actually happen, however, is unknown, as societal choices and technological change will determine future changes. This process will certainly be influenced by the availability of new scientific information, and its perception among the public and decision-makers. It is our hope that the FMAP Project will continue to provide such information into the oreseeable future.

Bibliography

Allen, G. R.: *Freshwater Fishes of Australia,* T.F.H. Publications, New Jersey, 1989

Archana Prabhakar: *Fish Immunology and Biotechnology,* Swastik Publications, Delhi, 2010.

Barman, R P : *Marine and Estuarine Fish Fauna of Orissa,* Zoological Survey of India, Delhi, 2007.

Barnum, Susan R.: *Biotechnology: An Introduction,* Belmont, Thomson/Brooks/ Cole, 2005.

Barrington, E. J. W.: *Biochemistry of Primitive Deuterostomians,* London, Academic Press, 1974.

Cahill, Lisa : *Genetics, Theology, and Ethics: An Interdisciplinary Conversation,* New York: Crossroad, 2005.

Chakraborty, Chiranjib : *Advances in Biochemistry and Biotechnology,* Daya, Delhi, 2005.

Daphne C. Elliott: *Biochemistry and Molecular Biology,* Oxford University Press, Delhi, 2005.

Dar, Ghulam Hassan : *Soil Microbiology and Biochemistry,* New India Publishing Agency, Delhi, 2010.

Farrand, John: *The Audubon Society Master Guide to Birding,* Alfred A. Knopf, New York. 1983

Featherly H. I.: *Taxonomic Terminology of the Higher Plants,* USA, Iowa State College Press, 1954.

Gotshall, D.W.: *Guide to Marine Invertebrates, Alaska to Baja California,* Sea Challengers. Monterey, California. 1994.

Hardy B.: *Biology and Agronomy of Forage Arachis,* Cali, International Centre for Tropical Agriculture, 1994.

Ivantsoff, W. : *Natural History of Dalhousie Springs.* W. F. South Australian Museum, Adelaide. 1989.

Jayasankar, P. and B. Anoop: *Identification of Marine Mammals of India*, Narendra Pub, Delhi, 2010.

Kamla Devi and D.V. Rao : *Poisonous and Venomous Fishes of Andaman Islands, Bay of Bengal,* Zoological Survey of India, 2003.

Kapoor, R.L. and M.L. Saini: *Plant Breeding and Crop Improvement*, CBS, Delhi, 1997.

Lake, J. S. : *Freshwater Fishes and Rivers of Australia*, Thomas Nelson Australia Ltd, Melbourne, 1971.

Larry V. McIntire : *Biotechnology: Science, Engineering, and Ethical Challenges for the Twenty-first Century,* Washington, DC: Joseph Henry Press, 1996.

Martin, K.C. : *Freshwater Fishes of the Northern Territory*. Northern Territory Museum of Arts and Sciences, Darwin. 1990.

Mitra, Abhijit and Kakoli Banerjee: *Marine Microbiology*, Narendra Pub, Delhi, 2004.

Narvekar, Raghunath : *Molecular Biochemistry : Principles and Practices,* Adhyayan Pub, Delhi, 2008.

Pillai, N.G.K.: *Marine Fisheries and Mariculture in India*, Narendra, Delhi, 2011.

Pollard, D. A. *Freshwater Fishes of South-eastern Australia*. McDowell, R. M. Reed, Sydney. 1980.

Ramakrishnan, T.V. : *Marine and Offshore Engineering*, Gene Tech Books, Delhi, 2007.

Razdan, Dheeraj: *Encyclopaedia of Marine Insurance : Its Principles and Practices*, Cyber Tech Pub, Delhi, 2011.

Sahoo, Dinabandhu and Prem Chand Pandey: *Advances in Marine and Antarctic Science*, APH, Delhi, 2002.

Saradha, T. : *Biology and Effects of Endosulfan and Dimethoate on Marine Molluscs*, Nidhi, Delhi, 2004.

Selvamani, B.R. and R.K. Mahadevan: *Marine Capture Fisheries*, Campus Books, Delhi, 2008.

Tomi Petr: *Fisheries in Irrigation Systems of Arid Asia*, Daya, Delhi, 2007.

Warren, E.B. and Herbert, R.A.: *Fishes of California and Western México. Pacific Marine Fishes,* T.F.H. Publications, Inc. Ltd. 1984.

Yadav, M : *Nutritional Biochemistry and Metabolism,* Arise Pub, Delhi, 2008.

Index

❑❑❑